COMBAT PAIR

THE EVOLUTION OF
AIR FORCE–NAVY INTEGRATION
IN STRIKE WARFARE

BENJAMIN S. LAMBETH

Prepared for the United States Air Force

Approved for public release; distribution unlimited

PROJECT AIR FORCE

The research described in this report was sponsored by the United States Air Force under Contract FA7014-06-C-0001. Further information may be obtained from the Strategic Planning Division, Directorate of Plans, Hq USAF.

Library of Congress Cataloging-in-Publication Data

Lambeth, Benjamin S.
    Combat pair : the evolution of Air Force-Navy integration in strike warfare / Benjamin S. Lambeth.
      p. cm.
    Includes bibliographical references.
    ISBN 978-0-8330-4209-5 (pbk,)
    1. Air warfare—United States—History. 2. Unified operations (Military science) 3. United States. Air Force. 4. United States. Navy—Aviation. 5. United States. Marine Corps—Aviation. I. Title.

UG633.L258 2007
358.4'24—dc22

2007044048

The RAND Corporation is a nonprofit research organization providing objective analysis and effective solutions that address the challenges facing the public and private sectors around the world. RAND's publications do not necessarily reflect the opinions of its research clients and sponsors.

RAND® is a registered trademark.

Cover design by Peter Soriano

Published 2007 by the RAND Corporation
1776 Main Street, P.O. Box 2138, Santa Monica, CA 90407-2138
1200 South Hayes Street, Arlington, VA 22202-5050
4570 Fifth Avenue, Suite 600, Pittsburgh, PA 15213-2665
RAND URL: http://www.rand.org/
To order RAND documents or to obtain additional information, contact
Distribution Services: Telephone: (310) 451-7002;
Fax: (310) 451-6915; Email: order@rand.org

Preface

This report was prepared as a contribution to a larger RAND-initiated study for the U.S. Air Force aimed at exploring new concepts for bringing land-based air power together with both naval aviation and surface and subsurface naval forces to enhance the nation's ability to negate or, if need be, defeat evolving threats in both major combat operations and irregular warfare. The report describes the evolution of Air Force and Navy integration in aerial strike warfare from the time of the Vietnam War, when any such integration was virtually nonexistent, to the contemporary era when Air Force and Navy air combat operations have moved ever closer to a point where they can be said to provide both a mature capability for near-seamless joint-force employment and a role model for other possible types of closer Air Force and Navy force integration in areas where the air and maritime operating domains intersect. It was sponsored by Major General R. Michael Worden, USAF, then-Director for Operational Plans and Joint Matters in the Office of the Deputy Chief of Staff for Air, Space and Information Operations, Plans, and Requirements (AF/A5X), Headquarters, United States Air Force. The research reported here was conducted within the Strategy and Doctrine Program of RAND Project AIR FORCE as a part of a fiscal year 2006 study titled "Exploring New Concepts for Joint Air-Naval Operations."

RAND Project AIR FORCE

RAND Project AIR FORCE (PAF), a division of the RAND Corporation, is the U.S. Air Force's federally funded research and development center for studies and analyses. PAF provides the Air Force with independent analyses of policy alternatives affecting the development, employment, combat readiness, and support of current and future aerospace forces. Research is conducted in four programs: Aerospace Force Development; Manpower, Personnel, and Training; Resource Management; and Strategy and Doctrine.

Additional information about PAF is available on our Web site at http://www.rand.org/paf/

Contents

Summary

During the more than three decades that have elapsed since the war in Vietnam ended, the U.S. Air Force and U.S. Navy have progressively developed a remarkable degree of harmony in the integrated conduct of aerial strike operations. That close harmony stands in sharp contrast to the situation that prevailed throughout most of the Cold War, when the two services lived and operated in wholly separate physical and conceptual worlds, had distinct and unique operating mindsets and cultures, and could claim no significant interoperability features to speak of. Once the unexpected demands of fighting a joint littoral war against Iraq in 1991 underscored the costs of that absence of interoperability, however, both the Air Force and the Navy quickly came to recognize and embrace the need to change their operating practices to accommodate the demise of the Soviet threat that had largely determined their previous approaches to warfare and to develop new ways of working with each other in the conduct of joint air operations to meet a new array of post–Cold War challenges around the world.

In the realm of equipment, the Navy in particular upgraded its precision-strike capability by fielding both new systems and improvements to existing systems that soon gave it a degree of flexibility that it had lacked throughout Operation Desert Storm, when its aviation assets were still largely configured to meet the very different demands of an open-ocean Soviet naval threat. Naval aviation also undertook measures to improve its command, control, and communications arrangements so that it could operate more freely with other joint air assets within the framework of an air tasking order (ATO), which by that time had become the established mission planning tool for large-scale air operations. Finally, in the realm of doctrine, there was an emergent Navy acceptance of the value of strategic air campaigns and the idea that naval air forces must become more influential players in them. For its part, the Air Force also embraced the new demands and opportunities for working more synergistically with its Navy counterparts both in peacetime training and in actual combat, where joint-force commanders stood to gain from the increased leverage that was promised by their working together more closely as a single team.

The single most influential factor that accounted for bringing the two services ever closer together in strike-warfare tactics, techniques, and procedures (TTPs) in this manner was the nation's ten-year experience with Operations Northern and Southern Watch, in which both Air Force land-based fighters and Navy carrier-based fighters jointly enforced the United Nations (UN)–imposed no-fly zones over northern and southern Iraq that had first been put into effect shortly after the conclusion of Operation Desert Storm. That steady-state aerial policing function turned out to be a real-world operations laboratory for the two services, and it ended up being the main crucible in which their gradual merger of operational cultures and styles was forged.

To be sure, despite this steady trend toward more harmonious Air Force–Navy cooperation, some lingering cultural disconnects between the two services persisted for a time throughout 1990s, most notably with respect to continued Navy discomfiture over having to operate within the framework of the Air Force–inspired ATO and the uneven way in which, at least in the view of many naval aviators, that mechanism made less than the most effective use of the nation's increasingly capable carrier-based forces. Nevertheless, the results of this steady process of integration were finally showcased by the near-seamless Air Force and Navy performance in their joint conduct of integrated strike operations in the largely air-centric war in Afghanistan in late 2001 and early 2002.

The uncommonly close meshing of land- and sea-based air involvement in that first round in the global war on terror, as well as the unprecedentedly prominent role the Navy played in the planning and conduct of the war, bore witness to a remarkable transformation that had taken place during the years since Desert Storm by way of a gradual convergence of Navy and Air Force thinking with respect to the integrated use of their air assets. Much energy was wasted during the early aftermath of Operation Enduring Freedom in parochial fencing between some Air Force and Navy partisans over which service deserved credit for having done the heavier lifting in the war, with Air Force advocates pointing to the preponderance of overall bomb tonnage dropped by the Air Force and with Navy proponents countering that it was carrier-based aircraft that flew the overwhelming majority of combat sorties. To say the least, that verbal sparring was completely unhelpful to a proper understanding of what integrated Air Force and Navy air operations actually did to produce such a quick allied win over the Taliban. At bottom, it remains an irrelevant toss-up as to which of the two services predominated in the precision-strike arena. Both brought indispensable combat capabilities to the joint effort. Any argument over whether Air Force or Navy air power was more important in achieving the successful outcome is tantamount to arguing over which blade in a pair of scissors is more important in cutting the paper.

The three-week campaign a year later to topple Saddam Hussein's regime in Iraq once again spotlighted the extent of operational integration that the two services had achieved in the conduct of joint air warfare since the first Gulf War of 1991. Operation Iraqi Freedom set a new record for close Navy involvement with the Air Force in the high-level planning and conduct of joint air operations. The five carrier air wings that took part in the campaign were better integrated into the ATO process than ever before, and the air war's deputy commander was a Navy two-star admiral. In all, the performance of Air Force and Navy strike assets in the first two American wars of the 21st century was replete with examples attesting to the giant strides that had been made in the integration of the two services' air warfare repertoires since Desert Storm. Both wars showed increased Air Force and Navy acceptance of effects-based thinking and planning, as well as a common use by the two services of the joint mission planning tools that had been developed over the previous decade and a half.

These real-world experiences suggest that the Air Force and naval aviation should now consider each other natural allies in the roles and resources arena, since they did not compete but rather mutually supported and reinforced one another in the achievement of joint strike-warfare goals. Indeed, when viewed from an operational rather than a bureaucratic perspective, the Air Force's and Navy's capabilities for air-delivered power projection are, and should be duly regarded as, complementary rather than competitive in the service of joint-force commanders, since land-based bombers and fighters and carrier-based fighters are not duplicative and redundant but rather offer overlapping and mutually reinforcing as well as unique capabilities for conducting joint warfare. Rather than continuing to engage in pointless either/or arguments over the relative merits of carrier versus land-based air power, Air Force and Navy proponents should instead be using their recent shared combat experience as a model for seeking ways to increase the synergy of their collective triad of long-range projection forces consisting of bombers, land-based fighters, and sea-based fighters that, taken as a whole, make up the nation's overall air power equation. (Figure S.1 graphically depicts this emergent synergy.)

By the candid admission of key leaders in both services, this process of integration in air warfare still has further headway to make before it will have realized its fullest potential. Nevertheless, it has advanced over the past decade and a half to a point where the air warfare arena is now by far the most developed realm of air-naval integration in the nation's joint-operations repertoire. Indeed, it constitutes an object lesson for the Air Force and Navy in the sorts of closer integration that can be successfully pursued by the two services in other mission areas where the air and maritime operating mediums intersect, as well as by the Air Force and Army in the air-land arena.

As for remaining areas where further work might be done by each service in the interest of closer air warfare integration, senior Air Force and Navy leaders have often cited continued communications problems and bandwidth-management shortcomings as one important set of challenges in need of continued attention. Another persistent sore spot between the Air Force and Navy, at least from the latter's perspective, concerns a rapidly looming problem in the electronic

Figure S.1
Attributes of Different Forms of Air Power

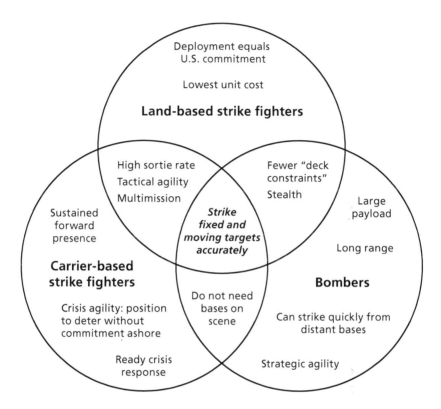

Deployment equals
U.S. commitment

Lowest unit cost

Land-based strike fighters

High sortie rate
Tactical agility
Multimission

Fewer "deck
constraints"

Stealth

*Strike
fixed and
moving targets
accurately*

Large
payload

Sustained
forward
presence

Long range

**Carrier-based
strike fighters**

Bombers

Do not need
bases on
scene

Crisis agility: position
to deter without
commitment ashore

Can strike quickly from
distant bases

Ready crisis
response

Strategic agility

RAND *MG655-S.1*

warfare mission area. When the Air Force decided to retire its aging EF-111 electronic jammer aircraft not long after Operation Desert Storm, the Navy and Marine Corps picked up the tactical electronic attack mission with their now greatly overworked EA-6B Prowlers, with the result that those aircraft became, to all intents and purposes, high-demand/low-density national assets. That arrangement has worked satisfactorily until now, but the EA-6Bs are rapidly running out of service life, the first replacement EA-18G Growlers will not enter fleet service until 2009, and the interservice memorandum of agreement that made the Navy the lead service in the provision of standoff jamming after Desert Storm expires in 2011. Accordingly, senior Navy leaders maintain that the Air Force will soon have to decide, conjointly with the Navy, what it intends to do by way of proceeding with timely gap-filler measures.

Still other possible joint ventures worth exploring in the training arena by the Air Force and Navy might include

- more recurrent exercises between the two services as focused instruments for spotlighting persistent cross-service friction points, to include greater Air Force involvement in Navy carrier air wing predeployment workups at Naval Air Station (NAS) Fallon and more Navy participation in Air Force Red Flag and other large-force training evolutions
- greater joint reliance on distributed mission simulation, which will entail high buy-in costs but can offer substantial long-term payoffs as fuel and associated training costs continue to soar
- a more holistic look at the joint use of training ranges, perhaps with a view toward ultimately evolving to a truly national range complex
- more comprehensive joint use of realistic adversary threats in training, not only in air but also in space and cyber operations
- extending integrated air warfare training to the surface and subsurface Navy
- enlisting the real-time involvement of air operations centers worldwide.

Many such initiatives are already being cooperatively pursued, or at least carefully considered, by the Air Force Warfare Center at Nellis Air Force Base (AFB) and the Naval Strike and Air Warfare Center at NAS Fallon, Nevada, with the primary limiting factor being insufficient funds to support them. As for additional areas of possible closer Air Force and Navy cooperation that pertain more to investments in equipment and hardware capability, the two services could usefully consider

- continued pursuit of ways of bringing their connectivity systems into closer horizontal integration

- greater attention to exploiting the promise of new electronic warfare means in joint warfare
- getting the greatest operational leverage for the least cost out of the high-commonality F-35 multirole combat aircraft that both services will be acquiring in the coming decade
- further coordination in setting agreed-on integration priorities.

Finally, a possibly high-payoff measure that would cost nothing beyond a determined Air Force and Navy effort to devote the right talent to it would entail a careful review of the accumulated base of documentation regarding all peacetime exercises and actual combat experiences of the two services over the past decade in search of identifiable friction points in integrated operations that were experienced and that may remain in need of attention and correction. Such an assessment could well illuminate previously unexplored areas of activity that could help move both services a step further toward achieving a fully mature joint strike-warfare repertoire.

Even with this much room remaining for further progress by the two services, however, the overall record of Air Force and Navy achievement in integrated strike-warfare planning and operations has been a resounding good-news story that is a credit to each service both separately and together. Air Force and Navy strike-warfare capabilities and repertoires have become almost seamlessly intertwined over the past three decades in a way, and to an extent, that cannot be said yet of any other two U.S. force elements. As such, they represent a role model for what can be done along similar lines elsewhere, not just in the interface between the air and maritime domains but even more so between the Air Force and Army when it comes to the most efficient conduct of joint air-land operations. Today, such commonality of purpose at the operational and tactical levels has become more important than ever as the nation finds itself increasingly reliant on the combined-arms potential that is now available to all services in principle for continuing to prosecute the global war on terror, while hedging also against future peer or near-peer competitors at a time of almost unprecedented lows in annual spending for force modernization.

Acknowledgments

I wish to express my thanks to numerous Air Force and Navy officers who so generously helped in one way or another to improve the content of this report during the course of its preparation. In early March 2006, I gained valuable insights into various aspects of evolving Air Force and Navy integration in joint warfare through conversations with Admiral Gary Roughead, USN, Commander, U.S. Pacific Fleet; Lieutenant General David Deptula, USAF, Commander, Kenney Warfighting Headquarters, Pacific Air Forces (PACAF), and members of his staff; Major General Dana Atkins, USAF, Director of Operations, U.S. Pacific Command; and Major Generals Loyd Utterback and Edward Rice, USAF, Deputy Commander and Director of Air and Space Operations, respectively, at Headquarters PACAF. I further benefited from a helpful early exchange on the subject of this report with Vice Admiral Lewis Crenshaw, Jr., USN, Deputy Chief of Naval Operations for Resources, Requirements, and Assessments, OPNAV N8.

I also wish to thank Vice Admiral James Zortman, USN, Commander, Naval Air Force, U.S. Pacific Fleet (COMNAVAIRPAC) and Commander, Naval Air Forces (CNAF), for kindly approving a three-day orientation visit by a group of RAND project staff to the aircraft carrier USS *John C. Stennis* (CVN-74) operating off the coast of Southern California on April 25–27, 2006, to observe carrier-qualification flight operations and to discuss various aspects of air-naval integration, as well as for warmly hosting our group for a breakfast discussion in his quarters at Naval Air Station (NAS) North Island, California, before our departure for the carrier. I am equally indebted to Rear Admiral David Buss, USN, at the time Commanding Officer of USS *John C. Stennis*, for freely sharing his time during our three-day visit despite a multitude of competing demands for his attention, and, in particular, for convening and chairing a most illuminating roundtable discussion on our study's behalf to explore strike-warfare issues and other facets of air-naval integration that also included his successor-designee as prospective Commanding Officer of *John C. Stennis*, Captain Brad Johanson; his Operations Officer, Captain Stephen Beckvonpeccoz; his Intelligence Officer, Commander Rick Stevenson; his Director of Strike Operations, Commander Mitch Hayes; Captain Vic Mercado, the Commodore of Destroyer Squadron 21 that was an accompanying part of the *Stennis* carrier strike group; and Captain Sterling Gilliam, USN, then-Deputy Commander (DCAG) of Carrier Air Wing 9 that was embarked at the time in *Stennis* for fleet carrier-qualification training. In addition, I express my gratitude to Captains Chuck Henry, Carroll Lefon, and Michael Manazir and to then-Commander James Bynum, USN, all on the COMNAVAIRPAC and CNAF staff, for collectively offering me two quality hours of their time after my return to North Island from *Stennis* to share their perspectives and experiences on the evolution of Air Force and Navy integration in air operations since Vietnam.

The analysis that follows has also been enriched by a number of instructive opportunities that I was privileged to have toward gaining a broad sampling of firsthand exposure to the world of Air Force and Navy air operations in the peacetime training environment in connection with RAND work on various air-warfare-related matters in years past. That exposure throughout much of the earlier formative history explored in this report included multiple sorties throughout the late 1970s and 1980s in the USAF's Nellis AFB range complex in F-100F, F-4C, and A-7K aircraft in various large-force training exercises, including four Red Flag evolutions, that included Navy and Marine Corps participation; six adversary training sorties in the TA-4J with VF-126 out of NAS Miramar, California, and NAS Fallon, Nevada, in 1980; three F-5F syllabus sorties with Navy Fighter Weapons School (TOPGUN) at Miramar in connection with my attending the first week of the TOPGUN course in 1980; two F-105F sorties later in 1980 in TOPGUN large-force training exercises that featured Air Force participation; four F-14A sorties, including two arrested landings in USS *Kitty Hawk* (CV-63), with VF-1 out of Miramar in 1983; a TA-7C sortie with the Naval Strike Warfare Center at Fallon in 1986; four air-to-air sorties in a Navy F/A-18B from VFA-125 out of NAS Lemoore, California, during the four-day Defensive Anti-Air Warfare Phase of the Weapons and Tactics Instructor (WTI) course offered quarterly by Marine Aviation Weapons and Tactics Squadron (MAWTS) One at MCAS Yuma, Arizona, in 1986; and an F-14A+ sortie with VF-24 out of Miramar in 1990.

More recently, such field orientation included completion of the USAF's week-long Combined Force Air Component Commander course at Maxwell AFB, Alabama, in 2002, which addressed Air Force–Navy integration issues in detail at the highest command levels, and a subsequent two-week visit to the Persian Gulf region in April 2007, which included several days in the Combined Air Operations Center of U.S. Central Command Air Forces (CENTAF) at Al Udeid Air Base, Qatar, and a 15-hour night E-3 combat-coded battle-management mission over Afghanistan, both of which offered multiple opportunities to observe mature Air Force–Navy integration in action on a daily basis. For the latter opportunity, I am indebted to the current CENTAF commander, Lieutenant General Gary North, USAF, who kindly invited me to accompany him on that trip.

Finally, for the many helpful comments and suggestions for improvement that they offered with regard to an earlier version of this report, I am grateful to Lieutenant General Deptula; Vice Admiral Evan Chanik, USN, then-Director, Force Structure, Resources, and Assessment (J-8), Joint Staff, and to his Executive Assistant, Captain Scott Craig, USN; Vice Admiral David Nichols, USN, Deputy Commander, U.S. Central Command and former Commander, Naval Strike and Air Warfare Center; Major General Stephen Goldfein, USAF, then-Commander, U.S. Air Force Warfare Center; Rear Admiral Thomas Kilcline, USN, Director, Warfare Integration and Assessment Division, OPNAV N8F; Rear Admiral Richard Gallagher, Director of Operations (J-3), U.S. European Command; Rear Admiral

Matthew Moffit, USN, Director, Fleet Readiness Division, OPNAV N43, and former Commander, Naval Strike and Air Warfare Center; Rear Admiral Mark Emerson, USN, Commander, Naval Strike and Air Warfare Center; Major General Michael Worden, USAF, Commander, U.S. Air Force Warfare Center; Brigadier General William Rew, Commander, 57th Wing, Nellis AFB, Nevada; Colonel William DelGrego, USAF, Lieutenant Colonel Drew Smith, USAF (Ret.), and Commander Terrence Doyle, USN (Ret.), Headquarters United States Air Force (AF/A5XS); Lieutenant Colonel Andrew Croft, USAF, Commander, F-15C Division, USAF Weapons School; and my colleagues Karl Mueller, David Shlapak, Alan Vick, and Lieutenant Commander Michele Poole, USN, the latter of whom served a year at RAND in 2005–2006 as a Navy Federal Executive Fellow. I also wish to thank my RAND colleague John Stillion and Vice Admiral John Mazach, USN (Ret.), former commander, Naval Air Force, U.S. Atlantic Fleet, and now Vice President for Business Development at Northrop Grumman Newport News for their incisive technical reviews of the final prepublication draft of this report.

Abbreviations

AAA	Antiaircraft Artillery
AB	Air Base
AEF	Air Expeditionary Force
AFB	Air Force Base
AGM	Air-to-Ground Missile
ATM	Air Tasking Message
ATO	Air Tasking Order
ATACMS	Army Tactical Missile System
ATFLIR	Advanced-Technology Forward-Looking Infrared
AWACS	Airborne Warning and Control System
BDA	Bomb Damage Assessment
C2	Command and Control
C4	Command, Control, Communications, and Computers
CAFMS	Computer-Aided Flight Management System
CAS	Close Air Support
CAOC	Combined Air Operations Center
CAP	Combat Air Patrol
CCIP	Continuously Computed Impact Point

CEC	Cooperative Engagement Capability
CENTAF	U.S. Central Command Air Forces
CENTCOM	U.S. Central Command
CFACC	Combined Force Air Component Commander
CFMCC	Combined Force Maritime Component Commander
CINCPAC	Commander in Chief for the Pacific
CNO	Chief of Naval Operations
COMNAVAIRPAC	Commander, Naval Air Force, U.S. Pacific Fleet
COMPTUEX	Composite Training and Underway Exercise
CONUS	Continental United States
CVW	Carrier Air Wing
DCGS	Distributed Common Ground System
ELINT	Electronic Intelligence
FAC	Forward Air Controller
GAT	Guidance, Apportionment, and Targeting
GBU	Guided Bomb Unit
GPS	Global Positioning System
HARM	High-Speed Antiradiation Missile
IADS	Integrated Air Defense System
IFF	Identification Friend or Foe
ISR	Intelligence, Surveillance, and Reconnaissance
JCS	Joint Chiefs of Staff
JDAM	Joint Direct Attack Munition
JFACC	Joint-Force Air Component Commander
JFCOM	Joint Forces Command

JSTARS	Joint Surveillance Target Attack Radar System
JTFEX	Joint Task Force Exercise
LANTIRN	Low-Altitude Navigation and Targeting Infrared for Night
LGB	Laser-Guided Bomb
MCAS	Marine Corps Air Station
MIDS	Multifunction Information Distribution System
MTI	Moving Target Indicator
NAS	Naval Air Station
NATO	North Atlantic Treaty Organization
NAVCENT	U.S. Central Command Naval Forces
NSAWC	Naval Strike and Air Warfare Center
OEF	Operation Enduring Freedom
RAF	Royal Air Force
RIO	Radar Intercept Officer
ROE	Rules of Engagement
ROK	Republic of Korea
SAM	Surface-to-Air Missile
SAR	Synthetic Aperture Radar
SEAD	Suppression of Enemy Air Defenses
SIGINT	Signals Intelligence
SIPRNet	Secure Internet Protocol Router Network
SLAM	Standoff Land Attack Missile
SLAM-ER	Extended-Range SLAM
SOF	Special Operations Forces
SOJ	Standoff Jammer
SPIN	Special Instruction
T3	Tomcat Tactical Targeting

T&R	Training and Readiness
TARPS	Tactical Air Reconnaissance Pod System
TEREC	Tactical Electronic Reconnaissance
TES	Tactical Exploitation System
TLAM	Tomahawk Land-Attack Missile
TST	Time-Sensitive Target
TTP	Tactics, Techniques, and Procedures
UAV	Unmanned Aerial Vehicle
UN	United Nations
UPT	Undergraduate Pilot Training
USA	U.S. Army
USAF	U.S. Air Force
USN	U.S. Navy
USNR	U.S. Naval Reserve
USS	United States Ship

CHAPTER ONE

Introduction

One of the most remarkable and praiseworthy features of American joint-force combat capability today is the close harmony that has steadily evolved over the past three decades in the integrated conduct of aerial strike operations by the U.S. Air Force and the U.S. Navy, along with the latter's closely associated Marine Corps air assets. This under-recognized and little-appreciated aspect of the nation's warfighting posture stands in marked contrast to the more familiar and contentious relationship between the two services in the roles and resources arena, where a fundamentally different incentive structure has tended to prevail and where seemingly zero-sum battles for limited defense dollars have appeared to be the natural order of things from one budget cycle to the next. As a former Air Force three-star general and fighter pilot recently remarked on this important point, although there remains "lots to be done at the budget table, tactically the [two] services are [now] bonded at the hip."[1] In a similar vein, a former commander of allied air forces in South Korea recently commented: "As the air component commander [in Korea], I don't differentiate between Air Force, Navy, [and] Marine Corps [contributions to the joint fight]. Joint . . .

---

[1] Email communication from Lieutenant General Tad Oelstrom, USAF (Ret.), Director, National Security Program, John F. Kennedy School of Government, Harvard University, June 1, 2006, commenting on Benjamin S. Lambeth, *American Carrier Air Power at the Dawn of a New Century,* Santa Monica, Calif.: RAND Corporation, MG-404-NAVY, 2005.

air power is crucial to success in this theater."[2] Indeed, in the words of a former Navy Fighter Weapons School instructor and now three-star commander of the Navy's Second Fleet, such integration "is now a part of the culture" of U.S. combat aircrews, regardless of whether the wings they wear on their uniforms are made of silver or gold.[3] In strong testimony to this fact, one today might easily encounter an Air Force F-15 or F-16 pilot, a Navy F/A-18 pilot, and a Marine Corps AV-8B pilot in an animated three-way conversation about strike-force employment tactics at Nellis Air Force Base (AFB), Nevada, Naval Air Station (NAS) Fallon, Nevada, or Marine Corps Air Station (MCAS) Yuma, Arizona, and be unable to tell which pilot was from which service without looking at the nametags and unit patches on their flight suits.

It has not always been that way. On the contrary, the Air Force and Navy have come a long way since the days of the Vietnam War and its early Cold War aftermath more than three decades ago, when the two services remained cultures apart, operated in wholly separate physical and conceptual worlds, and could claim no significant interoperability features to speak of. Once the unexpected demands of Operation Desert Storm in 1991 so starkly dramatized the downside consequences of that essential absence of interoperability between the two services, however, the Navy, in particular, responded to that wake-up call with all due alacrity and began implementing the many needed changes in its equipment, doctrine, and operating practices to accommodate the demise of its former open-ocean mission tasking against Soviet naval forces and its need to work more closely with its Air Force sister service in the conduct of joint air operations in dealing with littoral combat challenges around the world.

---

[2] Email communication from Lieutenant General Garry R. Trexler, USAF, then-Commander, Seventh Air Force; Deputy Commander, U.S. Forces in Korea; and Commander, Air Component Command, Republic of Korea/United States (ROK/U.S.) Combined Forces Command, June 1, 2006, also commenting on Lambeth, *American Carrier Air Power at the Dawn of a New Century.*

[3] Conversation with Vice Admiral Evan Chanik, U.S. Navy (USN), then-Director, Force Structure, Resources, and Assessment (J-8), the Joint Staff, Washington, D.C., August 1, 2006.

For its part, the Air Force likewise embraced not only the new demand for, but also the many new opportunities for, working more synergistically with its naval-aviator counterparts in both peacetime training and actual contingency operations in which both services stood to benefit from the enhanced performance promised by their working together more closely as a single team. From the most tentative initial stirrings of this early post-Vietnam move toward greater interoperability between the two services in the late 1970s, when Navy fighter and attack squadrons would periodically be invited to take part in the Air Force's recurrent Red Flag large-force training exercise conducted out of Nellis AFB, the two services registered ever-greater progress toward synchronized air operations throughout the 1990s, to a point where the fruits of that integration were finally realized during Operation Enduring Freedom over Afghanistan in late 2001 and further clinched by the all-but-seamless joint combat performance of the two services a year later during the three-week period of major combat in Operation Iraqi Freedom.

Although this process of operational integration between the Air Force and the Navy has not yet fully matured in light of its recognized but still-unrealized further potential and promise, its vector is now most definitely pointed in the right direction, in the almost unanimous view of senior leaders and line operators in both services. More important for the immediate purposes of this study, this steady trend toward ever-closer Air Force–Navy integration also has underscored the real synergy of American land- and sea-based strike forces when used wisely, in the right mix, and in the right joint-minded spirit. Indeed, it can now be said to have evolved to a point where it offers an object lesson in the kinds of integration that, with the application of the right thinking and effort by both services, can be achieved in other areas where U.S. air and maritime operations are likely to come together.

This report explores the evolution of that cooperative relationship in joint air warfare between the Air Force and Navy since the mid-1970s in terms of historical origins, new systems acquisition, the development of increasingly common tactics and operating practices, joint peacetime training, command and control, contingency operations, and actual combat experience. Its aim is to account not only for the many accomplishments that the two services have racked up to date, but also for unresolved issues and as-yet-unexplored ways of realizing further synergies between the two services. The growing synchronization of the two communities in air warfare has unfolded concurrently in three separate but related realms of activity—between the Air Force and Navy as separate uniformed services, between the air and maritime components in joint-force operations, and at the mission-area level, both within the air and space medium alone and in the broader domain where the air and maritime operating environments intersect. Some elements of this integration have had mainly to do with strategy, tactics, and concepts of operations. Others have related more to service-specific processes and procedures and to issues of institutional compatibility and interoperability. Illustrations from both categories will be presented in the discussion that follows, which will seek to tell the story of Air Force and Navy integration in air warfare as a casebook example of what the two services can yet accomplish by way of cross-service repertoire improvements more broadly defined.[4]

---

[4] The discussion above has intentionally used the terms *integration, synchronization,* and *interoperability* more or less interchangeably, since they all allude to a common process and dynamic. It bears noting, however, that these terms each have quite specific definitions and meanings in joint and service doctrine. For example, the standard Pentagon sourcebook on such matters defines integration as "the arrangement of military forces and their actions to create a force that operates by engaging as a whole"; interoperability as "the ability to operate in synergy in the execution of assigned tasks"; and synchronization as "the arrangement of military actions in time, space, and purpose to produce maximum relative combat power at a decisive place and time." All of these hair-splitting variations on a common theme apply in equal measure to the central topic of this report. See *Department of Defense Dictionary of Military and Associated Terms,* Department of Defense, Joint Publication 1-02, Washington, D.C., April 12, 2006, pp. 268, 277, and 524, respectively.

CHAPTER TWO

A Backdrop of Apartness

The first point to be stressed in any such assessment is that operational integration between the Air Force and the Navy is a fairly recent phenomenon in American military experience. For more than two centuries, the U.S. Navy was proudly accustomed to operating independently on the high seas, with a consequent need to be completely self-reliant and adaptable to rapidly changing circumstances far from the nation's shores and with the fewest possible constraints on its freedom of action. The nation's sea service was forward-deployed from the beginning of its existence and, throughout most of the Cold War, was the only service that was "out there" in and above the maritime commons and ready for action. Largely for that reason, until the second half of the 1980s, operations integration between the Navy and Air Force was not even a remote planning consideration. On the contrary, the main focus was rather on force *deconfliction* between the two services. Not only figuratively but also literally, the Air Force and the Navy conducted their daily routines in separate and distinct operating environments, and no operational synergies between the two services were produced—or even sought. Not surprisingly, a unique Navy culture and way of life emerged from this reality that set the Navy clearly apart from the Air Force's more structured and rule-governed way of conducting its missions.

The classic instance of this contrast in service styles was the war in Vietnam, in which different Air Force and Navy operating procedures essentially made integration between the two services in air warfare functionally impossible. At bottom, the main focus of the two services'

flight operations over both North and South Vietnam was simply stay-
ing out of each other's way.[1] The resultant fragmentation of air power
by operational-level Air Force and Navy commanders intent on pre-
serving their respective organic combat capabilities had the effect of
diminishing the overall efficiency of air operations by the two services,
thanks to a jury-rigged arrangement that one informed observer later
said "exemplified *disunity* of command."[2] As early as 1965, the mount-
ing daily sortie rate of Operation Rolling Thunder, as the initial American bombing campaign against North Vietnam was called, indicated
that better control was needed over the diverse American air assets that
were operating over that area. Up to that point in the still-nascent campaign, the four-star commander in chief for the Pacific (CINCPAC)
had delegated implementation authority for day-to-day Rolling Thunder planning to the commanders of the Air Force's Seventh Air Force
headquartered in Saigon, South Vietnam, and the Navy's Carrier Task
Force 77 deployed on Yankee Station off the coast of North Vietnam.
When that arrangement proved unsatisfactory, the Air Force and Navy
jointly developed the so-called Route Package system, in which North
Vietnam was divided into geographical areas numbered in sequence,
starting at the demilitarized zone and working northward. Mission
planners broke North Vietnam into seven regions. The Navy's Task
Force 77 got four of these, Route Packages II, III, IV, and VI-B adja-
cent to the coastline, since the carrier deck cycle and aircraft range
limitations made it easier for the Navy to operate on direct lines to the

---

[1] On this point, a former naval aviator who flew F-8s from the carrier USS *Oriskany* toward
the end of the Vietnam War recalled a hair-raising near-miss encounter when, thanks to
spotty cross-service information-sharing and consequent poor mission planning, a two-
plane section in which he was flying as a wingman was inadvertently dragged by his leader
directly underneath the flight path of an Air Force B-52 operating at a higher altitude on an
ARC LIGHT bombing mission over South Vietnam, with the result that a virtual hailstorm
of 750-lb bombs from the B-52 fell directly through his formation on all sides, barely avoiding an inflight fratricide incident. (Conversation with Captain Alan K. Steinbrecher, U.S.
Naval Reserve (USNR) (Ret.), San Marino, California, November 21, 2006.)

[2] Lieutenant General Phillip B. Davidson, U.S. Army (USA) (Ret.), *Vietnam at War: The
History, 1946–1975*, New York: Oxford University Press, 1988, p. 397.

littoral. The Air Force's Seventh Air Force drew Route Packages I, V, and VI-A.[3]

These "Route Packs," as they came to be called by aircrews and mission planners, were conceived as a part of Secretary of Defense Robert McNamara's graduated strategy of escalation, with successively more important and more heavily defended targets being those in the higher-numbered categories. Route Pack VI, embracing the northern-most part of North Vietnam, included the heavily defended Hanoi-Haiphong complex, the enemy's MiG fighter bases, and the greatest concentration of infrastructure targets. Because of its target density, it was divided into western and eastern halves as noted above, with the interior Package (VI-A) going to the Air Force and the littoral Package (VI-B) to the Navy.[4]

This fragmented approach to aerial warfare against North Vietnam stood in sharp contrast to the most basic beliefs held by Air Force airmen about the employment of air power dating back to World War II and before. Ever since the early admonishments of the outspoken Army air power advocate Brigadier General William "Billy" Mitchell, Air Force airmen had called for the centralized integration of all air assets under the control of one commander. Basic Air Force doctrine forged in World War II called for a single manager to orchestrate the use of air power most efficiently across the theater. Yet in Vietnam, the complex command arrangements that had been put in place made it impossible for senior leaders to establish a single manager for air operations. Because of the many differences between Air Force and Navy operating practices and procedures, a formal system of joint command and control was never established, and efforts to coordinate Air Force strikes out of Thailand with operations from Navy carriers in the Tonkin Gulf were accordingly rare. This failure denied the strike forces of the two services any opportunity to combine their capabilities

[3] Lieutenant Colonel Stephen J. McNamara, U.S. Air Force (USAF), *Air Power's Gordian Knot: Centralized versus Organic Control,* Maxwell AFB, Ala.: Air University Press, 1994, pp. 105–106.

[4] McNamara, *Air Power's Gordian Knot,* p. 45.

to greatest operational effect. To all intents and purposes, the Air Force and Navy fought separate air wars over North Vietnam.

These widely divergent service approaches to air operations persisted throughout the 1970s and early 1980s, with the final years of the Cold War after the nation's combat involvement in Vietnam ended in 1973 seeing little significant change from the previous pattern of segregated operations that were the norm throughout the eight-year air war in Southeast Asia.[5] Throughout those final Cold War years, the Navy's carrier battle groups figured most prominently in a sea-control strategy that was directed against Soviet naval forces, including long-range and highly capable shore-based naval air forces, in open-ocean engagements around the world. Because the Maritime Strategy of President Ronald Reagan's administration put the focus of American naval force projection more than a thousand miles away from the most likely focus of any Air Force combat operations in both Europe and the Pacific, such geographic separation, in an apt portrayal, "simply ruled out any concern with or interest in cross-service synergies at the operational or tactical levels."[6] Any combat operations against Soviet forces in the northernmost reaches of the Norwegian Sea or off the Kamchatka Peninsula in the Western Pacific would have involved solely the U.S. and Soviet navies, with no other force operations in the area. That accordingly freed the Navy to develop long-range fire-and-forget weapons such as the AIM-54 Phoenix air-to-air missile and the AGM-84D Harpoon

---

[5] One might cite as an exception to this general rule the case of Operation El Dorado Canyon, in which Air Force F-111s operating from the United Kingdom and Navy A-7s and F/A-18s from the carriers USS *Coral Sea* and USS *America* in the Mediterranean conducted punitive strikes against targets in Tripoli and Benghazi, Libya, in April 1986 in response to a Libyan-sponsored terrorist attack on a discotheque in West Berlin that killed two Americans and wounded more than 50. However, that operation was integrated more in appearance than in fact, considering that there was no face-to-face joint mission planning at the tactical level and that naval strike assets alone conducted the Benghazi portion of the attacks. At bottom, it was more an instance of Air Force and Navy aircraft conducting related but separate strike operations within the same battlespace. For further discussion, see Benjamin S. Lambeth, *The Transformation of American Air Power,* Ithaca, N.Y.: Cornell University Press, 2000, pp. 100–102.

[6] Major General John L. Barry, USAF, and James Blaker, "After the Storm: The Growing Convergence of the Air Force and Navy," *Naval War College Review,* Autumn 2001, p. 122.

antishipping missile that were unconstrained by any need for concern over the risk of fratricide or the possibility of causing unintended collateral damage should they go astray.

For its part, the Air Force was looking at a very different and more complex operating arena in which friendly and enemy aircraft would be simultaneously airborne and often commingled in the same block of airspace. Unlike the Navy, which was focused literally a thousand miles away—on the open-ocean environment, on the northern flank of the North Atlantic Treaty Organization (NATO) and the associated defense of northern Norway, and on Murmansk and the Kola Peninsula of the Soviet Union—the Air Force was preparing itself for joint operations in shared battlespace with the Army and with U.S. NATO allies in Central Europe. Given that stark dissimilarity in outlook and mission orientation, the Navy and Air Force, in a fair characterization, "simply thought about and operated within two separate conceptual worlds."[7]

As a result of these widely divergent operational mindsets and operating environments, a pronounced culture divide separated the Air Force and naval aviation in the strike-warfare arena. In telling testimony to this divide, Air Force pilots who participated in joint peacetime training exercises with their Navy counterparts during the early post-Vietnam years were often heard to tell horror stories about such (to them) seemingly cavalier and undisciplined Navy practices as last-minute unannounced changes in flight schedules, controlling agencies, radio frequencies, operating areas, and even mission profiles. For their part, Navy pilots who flew in similar joint training exercises routinely complained that the Air Force's allegedly overly rigid adherence to maintenance, operations, and crew-rest requirements greatly hampered its ability to be fully flexible in executing its assigned missions. One junior naval aviator in 1991 voiced a commonly heard refrain in this respect that neatly encapsulated the essence of the cultural divide from the Navy's perspective: "Naval aviators are fond of saying that Air Force pilots may only do something if it is written somewhere that they

---

[7] Barry and Blaker, "After the Storm," p. 122.

can, while Navy pilots may do whatever they want as long as it isn't written somewhere that they can't."[8]

With the passage, however, of the Defense Reorganization Act of 1986, more commonly known as the Goldwater-Nichols Act after its two sponsors, Senator Barry Goldwater of Arizona and Representative Bill Nichols of Alabama, the first halting steps toward closer Air Force and Navy integration in the peacetime training environment began to be taken, particularly in Europe and in the Mediterranean region, as that landmark legislation soon fostered ever-increasing jointness among the various individual service forces operating under the world-wide regional theater commanders in chief starting in 1987. These joint exercises soon offered clear hints at the interoperability challenges that the Air Force and Navy would confront in the immediate years to follow. In a particularly arresting early example of such hints, a Navy F-14A Tomcat accidentally downed an Air Force RF-4C Phantom II on September 22, 1987, during exercise Display Determination '87, one of the first of these embryonic attempts at closer joint Air Force and Navy aircrew training.

The RF-4 in question, based at Zweibrücken Air Base (AB), West Germany, was equipped with the then still-new tactical electronic reconnaissance (TEREC) system that had been designed to detect and geolocate land-based surface-to-air missile (SAM) radar emissions. For its part, the F-14 involved in the incident was operating from the carrier USS *Saratoga* in the central Mediterranean Sea. One goal of the exercise was to test the ability of the Air Force TEREC system to locate naval battle groups by their radar emissions. As the incident unfolded, the RF-4 was detected by the carrier battle group's surveillance radar while it was refueling from an Air Force KC-135 tanker. The F-14 was then vectored by its carrier-based controller to investigate the contact. The F-14 crew observed the RF-4 complete its refueling cycle and then initiate a descending turn away from the tanker. After receiving a radio call "Warning Red, weapons free," a standard exercise call at the time (as opposed to "weapons tight" or "weapons hold"), the inexperienced

---

[8] Lieutenant Dennis Palzkill, USN, "Making Interoperability Work," *Proceedings*, September 1991, p. 52.

junior-officer F-14 pilot was surprised by the call and asked his radar intercept officer (RIO) in the rear cockpit if he was supposed to shoot. The RIO's reply reportedly was: "Yeah, go ahead and shoot 'em."

To be sure, the RIO clearly had in mind making an exercise call, but his pilot was not on the same wavelength and accordingly armed his aircraft, selected an AIM-9L heat-seeking missile, and fired it at the RF-4. The first AIM-9L malfunctioned and failed to leave its launch rail. The F-14 then continued to close on the unsuspecting RF-4 from dead astern. Its pilot selected a second missile and fired again. This time the missile left the rail and hit the RF-4. Fortunately, the second missile had been launched inside its minimum range, so its warhead did not arm. However, it did knock the horizontal stabilizer off the RF-4, causing the aircraft immediately to depart from controlled flight. The two crewmembers promptly ejected successfully and were plucked from the water shortly thereafter by the carrier *Saratoga*, still completely in the dark as to why their aircraft had departed. They were convinced that their service careers were about to end abruptly before their very eyes until they learned that they had been inadvertently shot down by a friendly Navy fighter. Their first reaction was gratitude to know that they would not be saddled with the burden of having lost a multimillion-dollar aircraft to unknown causes. The incident was subsequently charged against the Navy's Class A aircraft mishap rate for fiscal year 1987.[9]

This revealing experience highlighted three important aspects of the Air Force's and Navy's first steps toward closer air-operations integration during the 1980s. First, it attested that the two services were indeed beginning to try to work more harmoniously than they had ever done before.[10] Second, it indicated that they still remained far short of

---

[9] I am indebted to my RAND colleague John Stillion for calling my attention to this sobering benchmark in U.S. military aviation history that fortunately ended happily for all. Dr. Stillion was serving as an active-duty Air Force RF-4 crewmember at Zweibrücken at the time the incident occurred. The above reconstruction is informed by his personal recollections of the incident, as well as by event-related operational details that can be found at http://www.tomcat-sunset.org/forums/index.php?topic=1160.0

[10] This increased cooperation, moreover, was becoming evident not just in the training environment but also in serious planning for major contingency operations in Central Europe.

that noble goal in some important respects. Third, and most tellingly, it offered dramatic proof that the operational costs of moving toward closer jointness could, at times, be remarkably high, even in peacetime. Nevertheless, the incident constituted a notable, if now long-forgotten, milestone in Air Force and Navy air warfare integration, the need for which was finally driven home most forcefully by the experience of both services in Operation Desert Storm just a little more than three years later.

As a former Air Force officer recalled from his experience as a NATO air operations planner at Headquarters Sixth Allied Tactical Air Force in West Germany: "When I arrived in 1989, there was limited joint and combined air participation in the existing planned air tasking order. We soon added more air forces from the Navy (and others) to COMSIXATAF's single air operations order [the acronym refers to Commander, Sixth Allied Tactical Air Force]. When we were done, there was a significant amount of Navy air, but more important, I believed that just enough capacity had been added by the Navy (and other NATO combined air forces) to enable a successful campaign." (Comments on an earlier draft by Lieutenant Colonel Drew Smith, USAF (Ret.), Headquarters United States Air Force, AF/A5XS, Washington, D.C., January 29, 2007.)

CHAPTER THREE

The Watershed of Desert Storm

Iraq's sudden and unexpected invasion of Kuwait in August 1990 presented naval aviation, in particular, with a new and unfamiliar set of challenges. Over the course of the six-week Persian Gulf War that began five and a half months later, the Navy's carrier force found itself obliged to surmount a multitude of new adjustment needs that only came to light for the first time during that campaign. Few of the challenges that were levied on naval aviation by that U.S.-led offensive, code-named Operation Desert Storm, bore much resemblance to the planning assumptions that underlay the Reagan administration's Maritime Strategy that had been created during the early 1980s to accommodate a very different set of concerns. Although naval aviators had routinely trained for and were wholly proficient at over-the-beach conventional strike operations, the Navy's carrier battle groups during that period were geared, first and foremost, to doing open-ocean battle against the Soviet Navy. As such, they were not optimally equipped for conducting littoral combat operations. They also were completely unaccustomed to operating within the Air Force's complex air tasking system for managing large-force operations involving 2,000 or more sorties a day that dominated the Desert Storm air war.

Simply put, the 1991 Gulf War in no way resembled the open-ocean battles that the Navy had planned and prepared for throughout the preceding two decades. To begin with, there were no opposed surface naval forces or enemy air threat to challenge the Navy's six carrier battle groups that participated in that war. Moreover, throughout the five-month buildup of forces in the region that preceded the war and the six weeks of fighting that ensued thereafter, the Navy did not operate independently, as was its familiar pattern throughout most of the Cold War, but rather in shared operating areas with the Air Force, Army, and Marine Corps.

During the initial planning workups before the start of Operation Desert Storm, some senior naval aviators sought for a time to push for a distribution of route packages between the Air Force and the Navy along familiar Vietnam-like lines. However, the designated joint-force air component commander (JFACC) for the impending campaign, Air Force then-Lieutenant General Charles Horner, rejected that proposal as an unacceptably suboptimal use of American air assets in joint warfare. Although Horner did not exercise formal *command* over the air assets of the Navy and Marine Corps that were deployed to the Gulf, he did wield operational *control* over them to an extent that empowered him to task them as he deemed appropriate to support the joint-force commander's air apportionment decisions. That arrangement was unprecedented in Navy experience. In the end, all four participating U.S. services came to accept, at least in principle, the need for a single jurisdiction over allied air power in Desert Storm. Yet three of them (not only the Navy but also the Marine Corps and Army) frequently chafed at the extent of authority given to General Horner to select targets and determine the details of flight operations.

Furthermore, the naval air capabilities that had been fielded and fine-tuned for open-ocean engagements, such as the extremely long-range (90-plus miles) Phoenix air-to-air missile carried by the F-14 fleet defense fighter, were of little relevance to the allied coalition's predominantly overland air combat needs.[1] Navy F-14s also were not assigned to the choicest combat air patrol (CAP) stations in Desert Storm because, having been equipped for the less-crowded outer air battle in defense of the carrier battle group, they lacked the redundant onboard target recognition systems that the rules of engagement

---

[1] James A. Winnefeld and Dana J. Johnson, *Joint Air Operations: Pursuit of Unity in Command and Control, 1942–1991,* Annapolis, Md.: Naval Institute Press, 1993, p. 115. See also Edward J. Marolda and Robert J. Schneider, Jr., *Sword and Shield: The United States Navy and the Persian Gulf War,* Annapolis, Md.: Naval Institute Press, 1998, pp. 180–181.

promulgated by U.S. Central Command (CENTCOM) required for the denser and more conflicted air operations environment over Iraq. Relatedly, because of the Navy's lack of a compatible command and control system that would enable receipt of the document electronically, the daily air tasking order (ATO) generated by the Air Force–dominated combined air operations center (CAOC) in Saudi Arabia had to be placed aboard two S-3 antisubmarine warfare aircraft in hard copy each day and flown to the six participating carriers so that the next day's air-wing flight schedules could be written. As for the Navy's other habit patterns and equipment items developed for open-ocean engagements, such as fire-and-forget AGM-84D Harpoon antiship missiles, ordnance supply planning purely to meet anticipated mission needs, and decentralized command and control, all were, in the words of the former vice chairman of the Joint Chiefs of Staff (JCS), Admiral William Owens, "either ruled out by the context of the battle or were ineffective in the confined littoral arena and the environmental complexities of the sea-land interface."[2] Naval aviation performed commendably in Desert Storm only because of its inherent professionalism and adaptability, not because its doctrine and weapons complement had been appropriate to the demands of the situation.

As examples of its deficiencies in equipment that impeded naval aviation's performance during the Gulf War, although it had clearly been equipped with advanced and capable combat aircraft by the time the war began, the Navy mainly dropped Vietnam-era unguided munitions, primarily the Mk 80-series 500-, 1,000-, and 2,000-lb general-purpose bombs. Throughout the war, the only carrier-based attack aircraft that was capable of self-designating laser-guided bombs (LGBs) was the A-6E. The A-7 and F/A-18 could also carry and deliver LGBs but only with the enabling support of nearby A-6Es that could laser-designate their targets for them, which was not an advisable tactic in heavily defended enemy airspace. Moreover, to remain safely above the enemy's man-portable infrared SAM and antiaircraft artillery (AAA) threat envelopes, they were required to operate solely from a stand-

---

[2] Vice Admiral William A. Owens, USN, "The Quest for Consensus," *Proceedings*, May 1994, p. 68.

off perch at medium and high altitudes, where visual weapon delivery techniques were less accurate because of the longer slant ranges to targets.[3] The Navy's electro-optically guided Walleye munition could be used only in daylight and in visual meteorological conditions. Carrier-based ground-attack aircraft also did not have anything like the weapons system video capability that was installed in the Air Force's F-111, F-117, F-15E, and F-16.[4]

Because of the Navy's lack of a significant precision-strike capability when its call to deploy for Desert Storm arose, its six carrier air wings that participated in the campaign were denied certain targets that were assigned to the Air Force instead by default. The participating carrier air wings also had to turn down some target-attack opportunities because of their lack of a penetrating munition such as the Air Force's Mk 84 improved 2,000-lb bomb. Other unmet Navy needs were for more LGBs, for automatic laser target designators for all strike aircraft, and for state-of-the-art mission recorders for conducting better bomb-damage assessment (BDA). One strike-fighter squadron's after-action report not long after the Gulf War ended remarked that the Navy's general lack of the sort of precision-attack capability that the Air Force had used to such telling effect in the war "was eloquent testimony that naval aviation had apparently missed an entire generation of weapons employment and development."[5]

---

[3] This was, of course, a problem for the Air Force's F-16s as well, which also predominantly dropped unguided free-fall bombs throughout Operation Desert Storm. Both the F-16 and F/A-18 were equipped with a continuously computed impact point (CCIP) conventional weapon aiming system, but that system offered a consistent near-precision unguided weapons delivery capability only at release altitudes of around 5,000 ft and below.

[4] "Navy to Boost Image, Relations with Press," *Aviation Week and Space Technology,* September 23, 1991, p. 55.

[5] Strike Fighter Squadron 87, "Aircraft—Yes, Tactics—Yes, Weapons—No," *Proceedings*, September 1991, p. 55.

CHAPTER FOUR

Post–Gulf War Navy Adjustments to New Demands

On balance, it would be hard to overstate the shock effect that the Desert Storm experience had on the Navy as a whole, to say nothing of its carrier air component, with respect to the newly emergent needs of joint strike warfare. As one rising naval aviator noted insightfully in 1992 in this regard: "Nearly two decades of narrow focus— on one-shot, small-scale, and largely single-service contingency operations—left naval aviation temperamentally, technically, and doctrinally unprepared for some key elements of a joint air campaign such as Desert Storm."[1] Admiral Owens put the point even more bluntly four years later: "For the Navy, more than any other service, Desert Storm was the midwife of change."[2] Unlike the Army and Air Force, in Admiral Owens's assessment, "the Navy left the first of the post–Cold War conflicts without a sense that its entering doctrine had been validated."[3] More specifically, he added, Desert Storm "confirmed the operational doctrines that the Army and Air Force had developed over the previous two decades, but it also demonstrated that the Maritime

---

[1] Commander James A. Winnefeld, Jr., USN, "It's Time for a Revival," *Proceedings*, September 1992, p. 34.

[2] Admiral William A. Owens, USN (Ret.), *High Seas: The Naval Passage to an Uncharted World*, Annapolis, Md.: Naval Institute Press, 1995, p. 4.

[3] Owens, "The Quest for Consensus," p. 68.

Strategy—the basic operational concept driving Navy planning since the early 1970s—did not fit the post–Cold War era."[4]

Fortunately, although naval aviation, like the Navy as a whole, entered the post–Cold War era ill-equipped for its new challenges and demands, the service quickly made the necessary adjustments in the wake of Operation Desert Storm. In the realm of equipment, largely in response to identified naval air shortcomings that were spotlighted by the Gulf War experience, the Navy stepped out smartly to upgrade its precision-strike capability by fielding both new systems and capability improvements to existing platforms that soon gave it a degree of flexibility that it had lacked throughout Desert Storm. First, it moved to convert its F-14 fleet defense fighter from a single-mission air-to-air platform into a true multimission aircraft through the incorporation of the Air Force–developed LANTIRN system, which allowed the aircraft to deliver laser-guided bombs both day and night.[5] Starting in 1997, the Navy ultimately modified 222 of its F-14s to carry the LANTIRN system, giving the aircraft a precision all-weather deep-attack capability that put it in the same league as the Air Force's versatile F-15E Strike Eagle. In the process, the F-14 relinquished much of its former strike escort role and left that to the F/A-18 with the AIM-120 advanced medium-range air-to-air missile as the Tomcat was transformed, in effect, into the deep precision-attack A-6E of old with its much-improved LANTIRN precision-targeting capability. (For its part, the venerable Intruder was facing imminent retirement after more than three decades of storied service with the fleet.)

The Navy leadership also freely acknowledged that its shortfall in precision-guided munitions had limited the effectiveness of naval air power in Desert Storm, a gap that it subsequently narrowed through the improvements to the F-14 noted above and by equipping more Navy and Marine Corps F/A-18s with the ability to fire the AGM-84E standoff land-attack missile (SLAM), to self-designate targets, and to deliver LGBs. It later acquired the extended-range SLAM-ER for the F/A-18, expanded the Navy's inventory of both LGBs and carrier-

---

[4] Owens, "The Quest for Consensus," p. 68.

[5] LANTIRN is an acronym for low-altitude navigation and targeting infrared for night.

based strike platforms capable of delivering them, and installed redundant positive target-identification capabilities on its air-superiority and multirole strike fighters.

To correct yet another equipment-related deficiency highlighted by the Desert Storm experience, naval aviation also undertook measures to improve its command, control, and communications arrangements so that it could operate more freely with other joint air assets within the framework of an ATO. Those measures most notably included gaining the long-needed ability to receive the daily ATO aboard ship electronically. In addition, the Navy made provisions for a more flexible mix of aircraft in a carrier air wing, which could now be tailored to meet the specific needs of a joint-force commander. The new look of naval aviation also featured a closer integration of Navy and Marine Corps air assets that went well beyond the mere "coordination" that had long been the rule hitherto. That initiative resulted in an unprecedented synergy of naval air forces occasioned by physically blending Marine F/A-18A and C squadrons into Navy carrier air wings as a matter of standard practice.[6]

Finally, in the realm of doctrine, there was an emergent Navy acceptance of the value of strategic air campaigns and the idea that naval air forces must become more influential players in them. As Admiral Owens noted as early as 1995, "the issue facing the nation's naval forces is not whether strategic bombardment theory is absolutely correct; it is how best to contribute to successful strategic bombardment campaigns."[7] In a major move to formalize and codify this new thinking, the Navy and Marine Corps on September 28, 1992, promulgated a fundamentally new strategy for the naval establishment in a white paper called . . . From the Sea.[8] That new mission orienta-

---

[6] The six land-based Marine two-seat F/A-18D all-weather strike-fighter squadrons were excluded from this integration because, with approximately 500 lb less internal fuel-carriage capacity occasioned by the space taken up by the aft cockpit for the second crewmember, that aircraft is operationally incompatible with single-seat Navy and Marine Corps A- and C-model Hornets.

[7] Owens, *High Seas*, p. 96.

[8] The full title of the 17-page document was . . . *From the Sea: Preparing the Naval Service for the 21st Century*, Washington, D.C.: Department of the Navy, 1992.

tion and vision statement put the main emphasis on power projection and explicitly envisioned naval forces as working jointly with both Air Force and Army elements to control events ashore.[9] It further recognized that for the near-term future, at least, the Navy's control of the high seas would remain uncontested and that the service's primary role in the post–Cold War era would be helping to enable joint-force operations in littoral areas. It also meant that, for the first time, the Navy would need to be thinking routinely in terms of integrated air operations with other services in the same battlespace. For example, the new vision, instead of focusing on defense against massed waves of Soviet bombers attempting to attack U.S. carriers on the high seas, focused more on joint warfare and associated connectivity programs, such as enabling Navy aircraft to work harmoniously with the Air Force's E-3 airborne warning and control system (AWACS) aircraft.

In essence, the new approach codified in . . . *From the Sea* reflected a fundamental shift from sea to land control, in which the underlying premise was that naval forces would now concentrate mainly on influencing events on a joint battlefield. A key part of that new vision was a change in emphasis toward greater Navy interoperability with the Air Force. As the new doctrine statement's title implied, the prevalent view was that naval forces would now operate *from* the sea rather than *on* it. Importantly in this respect, Admiral Owens stressed that "naval aviation must see itself as a component part of the full air power the nation can bring to bear on military problems, especially in support of land and air campaigns."[10] In keeping with this new notion of enabling, early post–Cold War naval thinking envisaged the nation's carrier air assets as the tip of the spear, in which forward-deployed naval forces would respond rapidly in crises, intervene to control escalation, and pave the way for the subsequent arrival of ground forces and land-based air power. Rear Admiral Philip Anselmo, the deputy director of space and electronic warfare on the Navy staff and a former fighter squadron

---

[9] On the origins and intent of that initiative, see the informed treatment by Captain Edward A. Smith, Jr., USN, "What '. . . From the Sea' Didn't Say," *Naval War College Review*, Winter 1995, pp. 9–33.

[10] Owens, *High Seas*, p. 49.

commander, well captured the early thinking that portrayed carrier air power in this manner, with carrier air wings conceived as being the first combatants on scene to enable the "war-winning" forces of the Army and Air Force.[11]

Finally, there were notable changes in naval aviation tactics, techniques, and procedures (TTPs) to make the nation's sea-based strike fighters and associated combat-support assets more compatible with the needs of joint warfare. After Operation Desert Storm, naval aviation's tactical emphasis shifted from air superiority and battle-group defense to multimission operations against possibly heavily defended targets ashore. That transition was greatly facilitated by the timely advent of the F/A-18 multirole fighter into the Navy's carrier air wings during the decade that preceded Desert Storm. Although the air-to-air skill set was retained, the focus of naval fighter training by the 1990s had taken a pronounced swing toward ground-attack operations, with a predominant stress on day and night precision strike.[12] The Navy's shift from an open-ocean to a littoral orientation inevitably made strike-fighter training more exacting and demanding, given that shift's accompaniment by new rules-of-engagement considerations, joint-service employment challenges, and the heretofore unprecedented demands of multimission tasking. The net effect of these improvements in equipment, doctrine, and concepts of operation was to transform U.S. carrier-based air power from a force configured mainly for sea control to one able to exploit sea control as a basis for enabling and participating in joint strike operations ashore.

As an essential precursor to what eventually became largely standardized Air Force and Navy strike-fighter tactics by the start of the 21st century, naval aviation leaders first had to impose a hitherto-elusive regime of standardization within their own strike-warfare community. During the early 1990s, in the immediate aftermath of Desert Storm,

---

[11] Edward J. Walsh, "The Copernican Revolution: U.S. Navy Leads the C4I Way for Joint Operations," *Armed Forces Journal International*, July 1995, p. 41.

[12] For an informed and lively account of introductory Navy F/A-18 strike-fighter training by a former Navy fighter pilot who immersed himself in that training for several weeks as a participant-observer, see Robert Gandt, *Bogeys and Bandits: The Making of a Fighter Pilot*, New York: Penguin Books, 1997.

there was widespread disagreement among both naval aviation leaders and line aircrews over the issue of whether fleet squadrons should train simply to meet the demands of the most likely threat or instead should train to the limits of their aircraft's capabilities in every mission category. Regarding this hotly debated question, an F-14 squadron executive officer wrote: "The various shore commands that are tasked with training our carrier battle groups have responded to the changing realities in a variety of ways; some aggressively and some hesitantly, some toward greater relevancy and some away from it, and some hardly at all. A single training philosophy has yet to emerge."[13]

In addition to the unresolved fleetwide question regarding the most appropriate focus for strike-fighter training, no single entity was responsible for overseeing carrier battle-group readiness at the squadron and air-wing level. During a typical West Coast air-wing turnaround in the mid-1990s, the notional "fingers in the pie" with respect to training oversight included, from the top down, Third Fleet; Naval Air Force, U.S. Pacific Fleet; Carrier Group One; Tactical Training Group, Pacific; the Naval Strike Warfare Center (or "Strike U," as it was more commonly known) at NAS Fallon; individual functional wings; and the battle group's own flag, air-wing commander, and squadron commanding officers.[14] In response to that perceived problem, the Chief of Naval Operations (CNO) moved to get the commanders of the Atlantic and Pacific Fleets to harmonize their respective battle-group and air-wing training programs more closely. Subsequent Navy-wide standardization initiatives made even further progress toward ameliorating the general absence of fleetwide common operating practices, with the result that training requirements and standards were finally codified in each wing's and squadron's operational readiness evaluations.

A major step forward by the Navy to further lock in this new stress on fleetwide tactics and procedures standardization came with the establishment in 1996 of the Naval Strike and Air Warfare Center (NSAWC) at NAS Fallon. In May of that year, the renowned Navy

---

[13] Commander J. D. Oliver III, USN, "To Train to Fight," *Proceedings*, September 1995, p. 40.

[14] Oliver, "To Train to Fight," p. 40.

Fighter Weapons School (more commonly known as TOPGUN) flew its last training sorties at NAS Miramar, California, as the Department of Defense's base realignment and closure process converted that facility (which, for decades, had been dubbed the Navy's "Fightertown") into a Marine Corps air station. At that point, advanced naval air-to-air training moved to Fallon as TOPGUN was integrated into NSAWC. "Strike U" for air-to-ground applications was already well established at Fallon, where it had been put in place more than a decade earlier at the instigation of then-Secretary of the Navy John Lehman in response to the flawed Navy performance in the Lebanon strike of December 1983 and to the fleet readiness deficiencies that that performance so starkly highlighted. The new NSAWC-to-be took under its roof at the now-superseded "Strike U" not only TOPGUN but also the Carrier Airborne Early Warning Weapons School (or "Top Dome," as it was known for short at the time), the E-2C community's advanced training program that likewise had previously been conducted at Miramar. Once those transfers were completed, NSAWC was formally established on July 11, 1996. In a single move, that consolidation capped a decade-long effort to bring together the various naval aviation communities into a common training arena and to standardize Navy strike planning and execution. It substantially bolstered the Navy's ongoing push toward attaining greater fleetwide commonality in air warfare training.

One of the first commanders of NSAWC, Rear Admiral Tim Beard, said that before its establishment, "we had factions throughout naval aviation—fighter, single-seat light-attack, and medium-attack, for example. And there was constant friction."[15] Strong leadership, however, finally prevailed in overcoming that parochialism. Instead of mission planning in which every faction got to play, operational needs for each mission began driving which aircraft, weapons, and force mixes were used. Said Admiral Beard: "If you were going in at night, there was only one airplane to use—the A-6. In the daytime, and if you didn't need max ordnance, the A-6 had no business going." This shift

[15] William B. Scott, "Fallon Becoming Navy's Air Combat 'Grad School,'" *Aviation Week and Space Technology,* March 8, 1999, p. 52.

in mission planning philosophy "served to get the one-team approach going. It did a lot to break down barriers."[16] One of NSAWC's early deputy commanders, Captain John Worthington, similarly recalled that "with the stroke of a pen, we took three strong legacies and organizations and turned them into one. The transition was difficult at first, but the organization has come together beautifully. As the primary authority on training, we can make sure everyone is teaching what we want them to teach."[17]

TOPGUN, now a newly constituent part of NSAWC, continued its air-to-air focus and retained its widely respected culture of world-class professionalism. Once the F/A-18 multirole fighter began arriving in the Navy's air wings, however, the scope of training concern at TOPGUN was expanded to include ground-attack operations. As noted above, there was already a well-established hard core of professionalism at "Strike U." Moving TOPGUN to NSAWC, in Admiral Beard's words, added significantly to that professionalism by "bringing in the razor's edge. Now, we play off the strengths of each [community]. We've broken down command barriers and put everybody together. Of course, nobody wants to give up their history and heritage. There's still some friction, but we're getting past that."[18] At bottom, the establishment of NSAWC finally gave naval aviation three centers of excellence under the aegis of a single command to provide integrated air-wing training for battle-group warfare. As a result, the warfighting potential of the Navy's carrier air wings and battle groups grew substantially compared to that of the typical battle group at the end of the Cold War a decade before.

In a final capstone measure to round out this steady progress toward total fleetwide standardization that had begun in the wake of Desert Storm, a new Navy Fleet Forces Command was activated in October 2001 under the command of Admiral Robert Natter just as

---

[16] Scott, "Fallon Becoming Navy's Air Combat 'Grad School,'" p. 52.

[17] Eric Hehs, "NSAWC," *Code One,* October 1998.

[18] Hehs, "NSAWC."

the Navy's role in the looming global war on terror was beginning to take initial form. That new command, which subsumed the U.S. Atlantic Fleet, had as its main goal the implementation of standard fleetwide practices aimed at achieving, in Admiral Natter's words, a more unified fleet that, while deploying from widely separated coasts, "embodies a shared and streamlined organization to complete the same training; executes common tactics, techniques, and procedures; and operates seamlessly around the world."[19] This new initiative, which was promulgated by the CNO at the time, Admiral Vern Clark, eliminated the previous arrangement whereby each of the three naval warfare communities (surface, submarine, and aviation) had separate but equal three-star commanders on each coast, an arrangement that led to often significant differences in standards and policies in each community on the two coasts. It changed that pattern fundamentally by providing for a three-star "lead type commander" on one coast and a two-star on the opposite coast, as well as for more detailed staff coordination between the three- and four-star fleet commanders on both coasts.

In the case of aviation, the commander of Naval Air Force, U.S. Pacific Fleet, then-Vice Admiral John Nathman, was designated the first lead type commander of the newly established Naval Air Forces worldwide. In the latter role, he became the principal advisor to Fleet Forces Command on aviation community issues, modernization needs, training initiatives, and operational concept development in the interest of fleetwide standardization. The main intent of that change was to make naval forces worldwide "as common as possible and unique only when required" by establishing uniform operating procedures, maintenance, training, and policies for executing community-specific programs, such as the management of allotted annual flying hours.[20] This new focus on forcewide readiness led to, among other things, the development of a more rigorous and uniform training and readiness (T&R) matrix for all aircraft types in the Navy's inventory and a consolidation

[19] Admiral Robert J. Natter, "New Command Unifies the Fleet," *Proceedings*, January 2002, p. 72.

[20] Natter, "New Command Unifies the Fleet," p. 73.

of both recurrent air-wing predeployment training and individual air-crew advanced weapons and tactics training at Fallon.[21]

Because the Soviet Union was well on its way toward dissolution by 1991 and, along with it, the long-standing Soviet air and naval threat to U.S. dominance of the high seas, the Navy would almost surely have been obliged to readjust itself to the new demands of the post–Cold War era sooner or later in any event, even had the Persian Gulf War never occurred. All the same, it was the unique demands that were levied on the Navy by Operation Desert Storm that were singularly most responsible for concentrating the Navy's attention on the need for a fundamental change in its equipment, doctrine, and concepts of operation and for the implementation of the many initiatives such as those outlined above that would prove essential for enabling the service's fullest possible participation as an equal player in future U.S. joint air operations.

---

[21] Conversation with Rear Admiral Thomas Kilcline, USN, Director of Air Warfare (N78), Office of the Chief of Naval Operations, Washington, D.C., October 5, 2004. The T&R matrix, it bears noting, establishes the training requirements and priorities for each aircraft type's primary mission areas, such as strike and antiair warfare, and provides type commanders a systematic framework for planning their training by specifying such matters as the relative importance of each mission event and how frequently it should be performed during a given training period. For amplification on this, see Appendix A, "U.S. Navy F/A-18 Fighter Training for Strike Missions," in John Schank, Harry Thie, Clifford Graf, Joseph Beel, and Jerry Solinger, *Finding the Right Balance: Simulator and Live Training for Navy Units,* Santa Monica, Calif.: RAND Corporation, MR-1441-NAVY, 2002, pp. 71–102.

CHAPTER FIVE

First Steps Toward Integrated Strike-Warfare Training

In keeping with its new post–Desert Storm operational orientation, the Navy undertook a number of positive initiatives during the early 1990s to expand its carrier air-wing training to include Air Force participation. One of the first practical applications of this changed Navy focus occurred in a joint exercise conducted in 1991 by the Air Force's 28th Bomb Wing and the Navy's Carrier Air Wing (CVW)-1. The air wing's after-action report observed: "Navy strike planners can create sanctuaries for B-1 strike using decoys, jamming and antiradiation missiles, while Navy aircraft are simultaneously striking targets in the same area." The report further noted that two B-1s added to a notional carrier air-wing strike package could increase the number of bombs on target by nearly 40 percent—at no cost to the Navy—since one B-1 can drop nearly as many Mk 82s as an entire F/A-18 squadron. After the exercise, the air-wing report called the B-1 "a valuable force multiplier."[1]

Similarly, in April 1993, a joint-force exercise called Kansas Global Lancer was conducted with the Air National Guard's 184th Bomb Wing at McConnell AFB, Kansas, and the air wing embarked in USS *Theodore Roosevelt*. Eight hours into the 23-hour exercise, bombers that had launched from McConnell joined up with a Navy strike package consisting of F-14s, F/A-18s, A-6Es, and EA-6Bs. Before the mission's

---

[1] After-Action Report, "28 WG/CAG-1 Joint Fallon Strike," December 31, 1991, p. 20, cited in Captain James W. Fryer, USAF, "Flying with the Bone," *Proceedings*, February 1995, p. 51.

scheduled time on target, Hornets, Intruders, and Prowlers conducted suppression of enemy air defense (SEAD) operations around the target area at a French-owned range on the eastern shore of Corsica. F/A-18s then swung to an opposing air role and established a CAP station over the western coast of Corsica. F-14s then escorted the B-1s into the defended airspace and engaged the Hornets while the B-1s blew through the ensuing melee unscathed to hit their assigned targets. The air wing's after-action report concluded that the "joint Navy-Air Force B-1B strike demonstrated maximum integration of strike, fighter, and search-and-destroy packages, significantly multiplying ordnance on target. . . . The strike was a total success and will provide a springboard for future real-world joint strike planning and execution."[2]

A closely comparable subsequent demonstration occurred during an early round of a recurrent joint interoperability training exercise called Roving Sands that was conducted in May 1994 in the White Sands area of New Mexico. The broader exercise of which this evolution was a part was a continuing effort directed by the chairman of the Joint Chiefs of Staff to explore joint theater air and missile defense and joint tactical air operations, with the principal players being Army air defense artillery brigades, Navy carrier air wings, Marine air control groups, Air Force theater air control system units, and a wide variety of operational flying units from all four services. This particular scenario featured a two-ship element of B-1s as the strike force, with attached escort provided by VF-211 F-14s and with VAQ-139 EA-6Bs and other CVW-9 aircraft providing detached cover against defending Air Force F-15s and F-16s and Army I-Hawk SAM sites. The exercise showed once again that interoperability between the participating units now worked well. It also prompted serious discussion of informally linking Air Force B-1 squadrons with Navy carrier air wings in a recurring arrangement in which the B-1s would train with the carrier air wings during the latter's predeployment workups in COMPTUEX (composite training and underway exercise) and JTFEX (joint task force exercise) evolutions, in subsequent strike-force training at NAS Fallon,

---

[2] After-Action Report on Kansas Global Lancer, "Joint USN-USAF B-1B Strike Postex," Command Sixth Fleet, April 10, 1993, p. 2, cited in Fryer, "Flying with the Bone," p. 51.

and during its actual deployment. An Air Force instructor pilot who participated in the exercise observed presciently: "We will probably be required to fight the first few weeks of our next war from the decks of our aircraft carriers and from the concrete runways of our U.S. bomber bases."[3]

In addition to the early examples of Kansas Global Lancer and Roving Sands, still another notable benchmark of this welcome trend was a major evolution called Rugged Nautilus '96, which was a joint-service exercise aimed at discouraging any possible terrorist challenges through a show of force in the Arabian Gulf while the 1996 Olympics were under way in Atlanta, Georgia. That joint exercise, which was conducted from July 1 to August 30, 1996, involved components from all four services that were forward-deployed in the Southwest Asian area of operations. Its primary goal was to test CENTCOM's ability to organize forces on short notice and to put in place the needed command and control arrangements to conduct air, ground, and surface naval operations.

The centerpiece of naval aviation's contribution to that exercise was CVW-14 embarked in USS *Carl Vinson*, which steamed with its accompanying battle-group complement through the Indian Ocean, transited the Strait of Hormuz, and assumed station in the Arabian Gulf in early July. With *Carl Vinson* the designated command and control ship, the Air Force, for the first time in its joint-operations experience, had to adjust to the novel experience of working with a *Navy* JFACC afloat. The Air Force's contribution entailed fighters, bombers, and tankers embodied in Air Expeditionary Force (AEF) III, which collectively flew 23 percent of the sorties in the largely Navy-centric exercise. In a typical AEF III contribution to Rugged Nautilus, which unfolded under the supervision of U.S. Central Command Naval Forces (NAVCENT), B-52s from the 2nd Bomb Wing at Barksdale AFB, Louisiana, flew nonstop from the continental United States (CONUS) to the Udari weapons range in Kuwait, where they each dropped 27 live 750-lb bombs in the first use of Air Force bombers in an overseas Navy-led theater training exercise. Other participating Air

---

[3] Fryer, "Flying with the Bone," p. 52.

Force units included the 4th Fighter Wing with F-15Es from Seymour-Johnson AFB, North Carolina; the 33rd Fighter Wing with F-15Cs from Eglin AFB, Florida; and the 20th Fighter Wing with F-16CJs from Shaw AFB, South Carolina. On multiple occasions, Navy F-14D Tomcats from CVW-14 provided armed escort protection for the Air Force B-52s as they flew over the Arabian Gulf.[4]

During the early 1990s, both services also showed an incipient doctrinal recognition of the need for closer integration in strike warfare. The Air Force's 1992 white paper called *Global Reach, Global Power* stated: "The Air Force and the Navy have both an opportunity and a responsibility to hone their cooperation for future operations [thanks to the] ever-increasing complementary interface between land- and sea-based air power."[5] Similarly, the Navy's revised 1994 white paper called *Forward . . . from the Sea* declared that "no single service embodies all the capabilities that are needed to respond to every situation and every threat."[6] To be sure, the two services continued to wage unrestrained knife fights in the roles and resources arena, as one would naturally expect of two organizations competing for limited funds in a highly charged political and bureaucratic environment. But steadily increasing integration was finally emerging between them at the operational and tactical levels of war.

Without a doubt the most sustained and influential factor in this bringing of the two services together in air warfare TTPs was the nation's ten-year experience of Operations Northern and Southern Watch, in which both Air Force land-based fighters and Navy carrier-

---

[4] I am grateful to Rear Admiral Thomas Kilcline, USN, Director of Warfare Integration and Assessment, OPNAV N8F, for bringing this exercise to my attention and for sharing his recollections of some of its highlights during a conversation in Washington, D.C., on August 1, 2006. For more on the Air Force's contribution, see William L. Dowdy, *Testing the Aerospace Expeditionary Force Concept: An Analysis of AEF's I–IV (1995–97) and the Way Ahead,* Maxwell AFB, Ala.: College of Aerospace Doctrine, Research, and Education, Air University, Research Paper 2000–01, 2000.

[5] *Global Reach, Global Power,* White Paper, Washington, D.C.: Department of the Air Force, December 1992, p. 6.

[6] John Dalton, Admiral Jeremy Boorda, and General Carl Mundy, Jr., "Forward . . . from the Sea," *Proceedings,* December 1994, p. 46.

based fighters jointly enforced the no-fly zones imposed by the United Nations (UN) over northern and southern Iraq that were first put into effect shortly after the conclusion of Operation Desert Storm.[7] That steady-state aerial policing function proved to be a true, real-world operations laboratory for the two services, and it ended up being the main crucible in which their integration in strike warfare was forged over time. By conscious choice, both services sent their best aircrews, tacticians, and intelligence officers to serve temporary-duty assignments in the supporting CAOCs in Turkey and Saudi Arabia to work together in the joint planning and execution of those nonstop air operations into and over Iraq. Over time, their working relations became more and more transparent and seamless. Once it became time, by the turn of the century, for the Air Force and Navy to start gearing up for Operation Enduring Freedom, both air warfare communities, thanks to their previous decade of having worked shoulder-to-shoulder in Operations Northern and Southern Watch, had become ever more comfortable and ever more accustomed to working together in a true joint-service partnership.[8] Viewed in hindsight, this convergence was not just a result of the Navy's objective need to get inside joint-force decision loops and to acquire the material wherewithal for remaining relevant in joint warfare but was even more a direct outgrowth of conscious senior leadership determination in *both* services, resulting in considerable part from their steadily evolved mutual trust and working relations over time, to move toward a more common operating culture across service lines when it came to coordinated joint-force execution.

---

[7] For a thorough and well-documented account of these joint and combined air operations, see Michael Knights, *Cradle of Conflict: Iraq and the Birth of the Modern U.S. Military,* Annapolis, Md.: Naval Institute Press, 2005, pp. 119–243.

[8] I would like to thank Vice Admiral Evan Chanik, USN, then-Director, Force Structure, Resources, and Assessment (J-8), the Joint Staff, for bringing this important insight to my attention during a conversation in Washington, D.C., August 1, 2006.

CHAPTER SIX

Continued Sources of Navy–Air Force Friction

Despite these nascent but salutary trends toward more harmonious cooperation in joint strike warfare, a number of cultural disconnects between the Air Force and the Navy persisted throughout the 1990s. One recurring manifestation of the cultural divide that still separated the two services in the air warfare arena came in the form of continued expressions of Navy discomfiture over the Air Force–inspired ATO and the way in which, at least in the view of many naval aviators, it sometimes made less than the best possible use of the nation's increasingly capable carrier-based strike forces. Ever since their first real exposure to operating in an ATO context during Operation Desert Storm, naval aviators had been inclined to chafe, sometimes quite insistently, at the alleged rigidity of that daily document that manages air combat operations and at its perceived insensitivity to certain unique features of sea-based air power, such as the inescapable operating requirements and limitations imposed by the carrier deck cycle.[1]

---

[1] As regards these special requirements, for every minute a carrier is headed into the wind during launch and recovery operations, it is committed to a predictable course that opponents can detect and track. It is thus deemed important as a matter of fleet tactics that launches and recoveries take place as rapidly as possible, consistent with due allowance for routine problems that inevitably arise from time to time in the course of such operations, such as a temporarily fouled deck and occasional aircraft "bolters," or failures to engage an arresting cable during attempted recoveries. So-called cyclic operations offer the best way to deal with these requirements. In such operations, a 1+0 cycle is one that lasts an hour from an aircraft's launch to its recovery. A 1+15 cycle lasts an hour and 15 minutes. In the instance of a notional 1+15 cycle, while one wave of aircraft is being launched, the preceding wave that was launched an hour and 15 minutes earlier will be holding overhead, with its pilots watching their constantly dwindling fuel levels and, as may be required, conducting recovery

For example, in the early aftermath of Operation Desert Storm, complaints were heard that in some cases, Navy squadrons had been tasked by the CAOC without adequate regard for their inherent constraints with respect to weapons loads, attainable aircraft ranges, and communications. Because of that perceived inadequate attention by the CAOC to important Navy-specific operational details, some Navy aircrews in Desert Storm simply rewrote their assigned mission profiles to their own satisfaction, to the understandable consternation of senior Air Force CAOC directors. (Cooperation with the CAOC by naval forces during the 1991 Gulf War with respect to ATO compliance was said to have varied from battle group to air wing to community and squadron.)[2] To be sure, the senior naval air liaison officer in the Air Force–dominated CAOC during Desert Storm was quick to acknowledge that the Navy's biggest problem with the ATO system had been its lack of the computer-aided flight management system (CAFMS)—an electronic ATO transmission system, which, as noted above, required that the ATO be flown in hard copy to each of the six participating air wings each day. Yet he still complained that "the 48-hour ATO cycle did not permit rapid response to mobile targets."[3]

Even the commanding general in charge of all Marine Corps aviation in Desert Storm complained that "the JFACC process of having

---

tanking near the carrier while the flight deck is being prepared for their recovery in sequence. In this manner, 20 to 30 aircraft can be kept airborne at any given time while the extra space thus freed up on the flight deck can be exploited for moving (or "respotting") aircraft to prepare for the next launch. During these gaps in flight operations, aircraft can also be moved back and forth to the hangar bay as may be required by the flight schedule or by maintenance needs. Within the span of a single deck cycle, whatever its duration, an air wing's aircraft are launched, recovered, de-armed, spotted, repaired, exchanged with hangar-deck aircraft, serviced, fueled, reconfigured with ordnance, and made ready for the next cycle. Managing the flight deck and respotting aircraft during such an intense operations flow is an exquisitely complex choreography in which the manipulation of assets and proper timing are absolutely crucial. For a fuller discussion of this complex process and the problems it can sometimes create for flexible large-force mission employment planning, see Peter Hunt, *Angles of Attack: An A-6 Intruder Pilot's War,* New York: Ballantine Books, 2002, pp. 53–55.

[2] Palzkill, "Making Interoperability Work," p. 50.

[3] Captain Lyle G. Bien, USN, "From the Strike Cell," *Proceedings*, June 1991 p. 59.

one single manager has its limitations. . . . It does not respond well to a quick-action battlefield. If you're trying to build a war for the next 72 to 96 hours, you can probably build a pretty good war. But if you're trying to fight a fluid battlefield like we were on, then you need a system that can react." He further complained about how the ATO process was "very cumbersome," with a daily document upward of 300 pages in length: "What I did to make it work for us—and I think the Navy did the same thing—was to write an ATO that would give us enough flexibility to do the job."[4]

Still another naval officer referred disparagingly to the "altar" of the ATO and how, in his view, it represented "the vast difference in world view between the U.S. Air Force and the U.S. Navy." He complained in particular about how "the people who published this tome never envisioned that a couple of junior enlisted air controllers on a three-week caffeine high in the back of a combat information center would have to flip through this six-pound chunk of fanfold paper on their knees to find the whereabouts of a tanker for their combat air patrol." As for the ATO's alleged inflexibility for supporting quick-response close air support (CAS), this A-6 bombardier-navigator complained that Air Force "reliance on advance planning, as illustrated by the air tasking order, does not allow for the often urgent and rapid sortie response needed during CAS."[5]

This persistent Navy discontent with the air tasking process, the latter of which was almost exclusively a mission-management artifact of the Air Force, was especially apparent throughout the contingency-response operations that were conducted by the Navy's carrier air wings, in conjunction with Air Force and allied air assets, over the Balkans throughout the 1990s. The first of those early joint contingency-response challenges was Operation Deliberate Force, the first serious test of American air power in the post–Cold War era that was prompted by a shelling attack by Bosnian Serbs against the city of

---

[4] Lieutenant General Royal N. Moore, Jr., USMC, "Marine Air: There When Needed," *Proceedings*, November 1991, p. 63.

[5] Lieutenant Commander Matthew J. Faletti, USN, "Close Air Support Must Be Joint," *Proceedings*, September 1994, p. 56.

Sarajevo in the former Yugoslavia that killed 38 innocent civilians. The purpose of the operation, a mini-campaign of coordinated NATO air strikes against selected Serbian targets in Bosnia-Herzegovina, was to deter further Serbian attacks against declared UN safe areas in Bosnia and to respond as necessary to any such attacks until they ceased.[6] Of the total number of sorties flown during this 11-day evolution from August 29 through September 12, 1995, the Air Force flew 774 (almost 52 percent), the Navy flew 583 (39 percent), and the Marine Corps flew 142 (about 10 percent). Of the 705 U.S.-conducted SEAD sorties, the Navy flew 395 (56 percent), the Air Force 244 (35 percent), and the Marine Corps 66 (9 percent).[7]

After Operation Deliberate Force was successfully concluded, there were recurrent expressions of Navy dissatisfaction over the Air Force's centralized control of mission tasking, particularly with respect to the air tasking message (ATM) that specified the type of munition to be used against particular targets. In an unavoidable policy decision that largely prompted these Navy complaints about alleged Air Force "inflexibility," the commander of allied air forces for NATO's southern command, Air Force then-Lieutenant General Michael Ryan, personally selected each target and individual aimpoint attacked by all Deliberate Force aircrews out of his and his uniformed and civilian superiors' acute concern over the adverse political consequences that could be incurred by either accidental fratricide or collateral damage and noncombatant casualties. That direct involvement on his part inescapably introduced an element of delay into the process of issuing and managing the daily ATM, which naturally, in turn, affected the tanker flow and other prebriefed events whenever a change in the ATM

---

[6] The most thorough available treatment of the operation is the final report of Air University's Balkans Air Campaign Study edited by Colonel Robert C. Owen, USAF, *Deliberate Force: A Case Study in Effective Air Campaigning,* Maxwell AFB, Ala.: Air University Press, 1999.

[7] Lieutenant Colonel Richard L. Sargent, USAF, "Deliberate Force Combat Air Assessments," in Owen, *Deliberate Force,* p. 346. It is worth noting here that almost half of the combat sorties flown were SEAD sorties, a fact that points up the extremely low risk tolerance of U.S. decisionmakers.

cycle was introduced.[8] As just one example, shortly before one scheduled target attack, the CAOC received notification that a company of French peacekeepers might be dangerously close to the target. Instead of canceling the mission outright, the CAOC retasked it for another target within the same planning cycle. That change then rippled into the tanker flow plan and into the tasking for all other assets that had initially been directed to support the strike operation. Such incidents led to a near-constant state of change in the tasking of General Ryan's air assets, including those operated by the Navy, and to a compressing of the original 24-hour planning cycle as a result of major modifications in the ATM and master air attack plan, sometimes occasioning understandable Navy expressions of concern that naval strike assets were not being optimally employed.[9]

---

[8] Colonel Christopher M. Campbell, USAF, "The Deliberate Force Air Campaign Plan," in Owen, *Deliberate Force*, p. 111.

[9] With regard to some earlier air activities over the Balkans that ended up being a warm-up exercise for this contingency response, an F/A-18 squadron commander who participated in those peacekeeping operations said of the ATM used by the CAOC that it was "complex, slow (24-, 48-, and 72-hour decision cycle times), rigid (little adjustment possible), useful as a means of coordinating aircraft movement—but little more. It engendered a passive mentality and internal friction by dictating tactics to the tacticians and integrating targeting, scheduling, and intelligence at the top rather than at the bottom." This Navy critic further charged that the ATM, which insisted on the type and quantity of ordnance to be carried, was grounded fundamentally on mistrust of aviators on the part of senior CAOC leaders, as in: "We'd better tell these wild pilots exactly what to carry, where to carry it, and how to use it . . . or else they'll mess it up." He denigrated the mechanism as "the largest obstacle to effective tactical air operations in support of ground forces in Bosnia. It focused on quantity instead of quality, it reflected a mechanistic, almost brutish approach to warfare—hardly desirable when dealing with an agile foe." Fairly enough, he conceded one of the ATM's most crucially important functions, namely, "scheduling the movement and rendezvous of large numbers of aircraft." Yet, he complained, "it had little capacity to put the right aircraft with the right pilot at the right place at the right time." Instead, "it spewed forth pounds of written text that coincided only by chance with events on the battlefield." While respecting those in the CAOC who had worked so hard at building and refining the ATM, he suggested that the problem "lies in the thinking that designed and continues to defend [it]." He further faulted the alleged "assumption that battlefield air operations can be segmented into neat and tidy 24-, 48-, or 72-hour segments that will permit the theater commander to closely—if not personally—control tactics by micromanaging individual aircraft and pilots." (Commander Daniel E. Moore, USN, "Bosnia, Tanks, and 'From the Sea,'" *Proceedings*, December 1994, p. 42.)

In light of the abundance of strike assets (more than 280 Air Force, Navy, Marine Corps, and allied aircraft) that were available for Operation Deliberate Force and of the fairly limited target set that they were being tasked to service, it was easy in hindsight to see how some people, notably including U.S. carrier-based naval aviators, outside the heart of the operation might wonder why it was so difficult to stick to a tasking cycle with a minimum number of changes and disruption. As a subsequent Air Force assessment frankly put it, "the compressed ATM cycle . . . rankled many people outside the CAOC." Yet be that as it may, this same assessment continued, "the CAOC staff had a clear understanding of the situation driving the ATM cycle. . . . Navy Captain [Kenneth] Calisle, the deputy chief of plans, agreed that planning inside the ATM cycle, together with hardware problems associated with disseminating the final product, presented a challenge to everyone involved. Nevertheless, he could see no other way to react to the commander's guidance."[10]

There also were recurrent flashes of rivalry between the two services over which aircraft types and which service should fly particular missions, simply because everyone wanted to be a player to the fullest degree possible. There were problems as well with communications interoperability, in particular with the CAOC and the Air Force's airborne command and control center getting through to USS *Theodore Roosevelt* and USS *America* in the Adriatic when the carriers operated together, which put a heavy burden on the limited satellite bandwidth available. Although none of these cultural differences seriously hampered the course and outcome of the operation, they showed, as one analyst put it, that "despite more than a decade of 'purple' experience, joint operations [remained] far from seamless and need further attention prior to future conflicts."[11]

Some of the complaints indicated above, especially from the more junior naval aviators, merely reflected a less than complete understanding of the CAOC's air tasking process and what lay behind it, plus the

---

[10] Mark Conversino, "Executing Deliberate Force: 30 August–14 September 1995," in Owen, *Deliberate Force,* p. 161.

[11] Sargent, "Deliberate Force Combat Air Assessments," in Owen, *Deliberate Force,* p. 336.

fact that until Desert Storm and the subsequent contingency operations over the Balkans and Southwest Asia in the 1990s, the naval air warfare community had never really been exposed to the ATO as a mission management tool.[12] It was not as though these young naval aviators were somehow retrograde in their thinking or simply failed to "get it." Quite to the contrary, those experienced carrier-based pilots and naval flight officers who voiced the criticisms and complaints laid out above were no less professionally able and tactically astute than their Air Force brethren. They merely had been brought up in a peacetime training environment that had been governed and dominated by far more easily manageable single air-wing operations.

Furthermore, particularly in the case of air operations over the Balkans during the 1990s, most of the expressed Navy dissatisfaction with and complaints about the ATM system would have existed under just about any alternative mission management arrangements as well, including the Route Pack system that was employed over North Vietnam nearly a generation before. The fact is that NATO air operations over the former Yugoslavia and the concurrent coalition sorties flown in Operations Northern and Southern Watch were instances of highly constrained force employment, in which it was not possible for CAOC planners to make optimal use of *any* military assets, Navy or any other. In those cases, the ATM served, in the main, as a convenient lightning rod for Navy complaints that were actually traceable instead to the severe operating limitations that were imposed on senior commanders by U.S. political leaders in the interest of avoiding fratricide, collateral damage, noncombatant civilian casualties, and other violations of standing rules of engagement (ROE), with the intent both to reassure nervous and reluctant NATO allies and to prevent tactical mis-

---

[12] To expand on this point, the ATO does indeed, of necessity, detail specific aircraft to specific missions, areas, and times and indeed requires 48 to 72 hours to plan starting from scratch because of the many concurrent aircraft movements it seeks to schedule and deconflict. Yet specific changes in the ATO's execution can be made and carried out in mere minutes as the needs of the moment may dictate. More to the point, the ATO schedules aircraft to be where and when planners anticipate they will be needed, but it does not predetermine the targets of those aircraft in every case, particularly now that time-sensitive target (TST) attacks have become the rule rather than the exception.

takes from producing undesirable strategic consequences. More to the point, most of the Navy complaining about the Air Force–dominated air tasking process throughout the 1990s in both the Balkans and over Iraq emanated from outspoken working-level officers who erroneously faulted the ATO, the CAOC, and the Air Force in pretty much equal and indiscriminate measure for perceived problems that actually emanated from such related but separate causal factors as ROE and special-instructions (SPINs) constraints, top-down political direction of targeting, heightened leadership concern for avoiding untoward incidents, and other sources of friction in execution that had nothing to do whatever with the air tasking process per se.[13]

The more senior naval aviation leaders, in contrast, tended to take a more understanding view of the ATO and of its advantages as well as its drawbacks in successfully managing large-force air operations. For example, shortly after Operation Desert Storm, the director of air warfare in the office of the CNO, Rear Admiral Riley Mixon, said of the ATO that the six-month experience of Operation Desert Shield had enabled U.S. naval forces in the Red Sea and Persian Gulf to learn how to operate within that system, adding: "I do not know of a better way to orchestrate 2000–3000 sorties per day from the four services and the numerous allied forces participating."[14] Another naval officer writing in the same spirit a year later stressed the need highlighted by Desert Storm for the Navy to "learn more about the joint process. The more we learn about the ATO system and how it functions, the better we can operate within its confines and make it meet Navy requirements." He acknowledged that "it seems likely that the Air Force users simply receive the message directly into a software management system, with which they can process the vast amount of data painlessly." He added: "Our collective experience with the Air Force's way of managing air-

---

[13] Vice Admiral David Nichols, USN, the deputy commander of CENTCOM and deputy Combined Force Air Component Commander (CFACC) during both Operation Enduring Freedom and the major combat phase of Operation Iraqi Freedom, was adamant on this point in offering critical feedback on an earlier version of this report. (Conversation with VADM Nichols at CENTCOM headquarters, MacDill AFB, Florida, February 7, 2007.)

[14] Rear Admiral Riley D. Mixon, USN, "Where We Must Do Better," *Proceedings*, August 1991, p. 39.

borne assets during war" had the effect of being a needed wake-up call. "Our traditional (and usually appropriate) reluctance to sign up to the other services' ways of doing business caught us short, and we were unprepared to offer an alternative."[15]

Also, it must be admitted, the cumbersome format of the early-generation ATO that was used by the CAOC in Desert Storm was, in many ways, an unimproved carryover from NATO's Cold War requirement for managing and deconflicting massed offensive and defensive air operations against Soviet and Warsaw Pact forces in Central Europe. As such, it was anything but user-friendly even for Air Force aircrews, let alone for Navy and Marine Corps aviators who were completely unaccustomed to working within its framework. Even today, the more streamlined air tasking process that has been rendered so much more adaptable thanks to subsequent improvements in command and control and information technology remains an imperfect device for the seamless conduct of joint and combined air warfare. That said, either it or something like it was and remains indispensable for preventing aerial friendly fire engagements and otherwise choreographing the complex flow of theaterwide air operations on a scale vastly larger than those of any single carrier air-wing strike package.[16] What matters most for the purposes of this discussion is that Air Force and Navy strike planners have shown remarkable progress over the past decade toward making the air tasking process more efficient and mutually agreeable. A big part of the explanation for this has been the revolution that has occurred since Desert Storm in getting senior naval involvement in the CAOC all the way up to the highest command level, as well as rank and file naval-aviator participation in ever larger numbers not just as air-wing representatives and weapon system subject-matter experts, but as full-

---

[15] Lieutenant Commander Larry Di Rita, USN, "Exocets, Air Traffic, and the Air Tasking Order," *Proceedings*, August 1992, pp. 62–63.

[16] For example, the ATO assigned Mode II identification friend or foe (IFF) codes to all missions and included the inflight refueling schedule for all Air Force and Marine Corps tankers. Another naval officer acknowledged that thanks to the ATO-dominated command and control system that was used, more than 65,000 aircraft sorties were handled without any midair collisions or blue-on-blue engagements. (Kevin E. Pollack, "Desert Storm Taught Us Something," *Proceedings*, January 1995, p. 68.)

fledged members of the CAOC battle staff able to weigh in and effect meaningful changes where needed.

NATO's Operation Allied Force against Serbia in the spring of 1999, the second major contingency-response air war of the post–Desert Storm era, saw considerably closer integration of naval air assets in allied strike operations, including the first combat use of the F-14D "Bombcat" from VF-41 embarked in USS *Theodore Roosevelt*. That experience showed convincingly that naval aviation was now capable of assuming a considerably larger role against time-critical targets. Thanks to their onboard LANTIRN infrared-imaging targeting pod, F-14 aircrews were able to contribute significantly as airborne forward air controllers, which allowed less-capable F/A-18s to hit positively identified targets. Although the 74 aircraft in CVW-8 embarked in *Theodore Roosevelt* accounted for only 4,270 out of a total of 38,000 combat sorties flown by all NATO aircraft, they made an important contribution to Operation Allied Force all the same by effectively servicing 447 tactical targets and 88 fixed targets in Serbia and Kosovo. The battle group commander, Real Admiral William Copeland, later reported that his sea-based air assets had been "fully integrated" with other allied operations and had sustained unusually high sortie rates as his aircrews were tasked to fly double- and triple-cycle combat missions each day and night.[17]

As in the earlier case of Operation Deliberate Force, there were complaints from some Navy quarters that the best use was not always made of certain naval systems that were available in principle to the air component commander. For example, the Sixth Fleet's battle staff consistently felt that its air wing was improperly treated by the CAOC as merely another allied fighter squadron rather than as the integrated strike force with intelligence, surveillance, and reconnaissance (ISR) and command and control backup that it actually was. Navy operators also pressed repeatedly to have the F-14's tactical air reconnaissance pod

---

[17] Gordon I. Peterson, "Naval Aviation Spearheads Operation Noble Anvil," *Sea Power*, June 1999, p. 1. Double- and triple-cycle missions entail sea-based sorties that are two and three times the duration of a normal single 1+0 or 1+15 cycle (enabled by multiple inflight tanker hook-ups as needed) because of extended distances to target or other mission-specific demands, such as the need to hold over a target area for providing on-call CAS.

system (TARPS) used for direct mission support, whereas the CAOC tended to persist in using it primarily for supporting BDA.[18]

Also, for a time throughout the 1990s, serious thought was given in naval aviation leadership circles to the possibility that at least in some littoral scenarios, a Navy JFACC might serve aboard a carrier and perform the same sorts of command functions that General Horner had fulfilled from his Desert Storm air operations center in Saudi Arabia. After years of arguing that a sea-based JFACC "could not only manage moderately sized air operations, it also would be more agile, responsive, and better able to deal with mobile and time-critical targets," however, senior naval aviation leaders finally accepted the hard reality that the demands on an air operations center for other than single-shot demonstrative or punitive strikes were such that they could not, as a rule, realistically expect to plan and conduct significant air operations from forward-deployed carriers. In 2000, the Navy's JFACC coordination committee conceded that owing to deck-space limitations and other constraints, a carrier can readily support only a maritime ATO, namely, the carrier's own flight operations.[19] That admission was in no way a preemptive surrender to Air Force operating preferences but simply a candid Navy recognition and acceptance of a tangible joint-operations reality. According to the Navy's JFACC committee, a true sea-based air operations center would, at a minimum, require a fleet command ship like USS *LaSalle* or USS *Mount Whitney*, ships that are not routinely deployed with carrier battle groups and that could require weeks to be surged from the United States or from their home ports abroad.

Navy leaders also later freely acknowledged that Operation Allied Force did not represent a good illustration of the ability of naval air forces to deploy rapidly and to remain on station indefinitely, since the profusion of land bases in Italy and elsewhere so close to the war zone gave the Air Force and Marine Corps all the ramp space they needed. In that operation, carrier-based Navy aircraft and shore-based Marine

---

[18] Conversation with Vice Admiral Daniel J. Murphy, USN, Sixth Fleet commander, aboard USS *LaSalle*, Gaeta, Italy, June 8, 2000.

[19] Lieutenant Commander Steve Rowe, USNR, "Saving Naval Aviation," *Proceedings*, September 2000, p. 34.

jets played only a minor role, contributing only 5 percent of the air-craft compared to 54 percent by the Air Force and 41 percent by the participating NATO allies.[20] It would take Operation Enduring Freedom over Afghanistan two years later to spotlight the indispensability of the nation's carrier forces in situations in which needed host-nation access for land-based fighter operations was simply too far away from the objective area to enable such operations on a routine basis.

---

[20] Benjamin S. Lambeth, *NATO's Air War for Kosovo: A Strategic and Operational Assessment,* Santa Monica, Calif.: RAND Corporation, MR-1365-AF, 2001, p. 33.

CHAPTER SEVEN

A Convergence of Integration over Afghanistan

The attacks planned and executed against the United States by Osama bin Laden and his al Qaeda terrorist organization on September 11, 2001, not only confronted all of the U.S. services with a no-notice call to arms, they levied on the nation a demand for a deep-attack capability in the remotest part of Southwest Asia where the United States maintained virtually no access to forward land bases. That unusual demand required that the Navy's carrier force step into the breach by providing the bulk of strike-fighter participation in the joint air war over Afghanistan that ensued soon thereafter.[1]

To be sure, Air Force heavy bombers also played a prominent part in that air-centric campaign, code-named Operation Enduring Freedom (OEF), by flying from the British island base of Diego Garcia in the Indian Ocean and, in the case of the B-2 stealth bomber (which flew six missions against the air defenses of the ruling Taliban during the campaign's first two nights), all the way from Whiteman AFB, Missouri, and back. Indeed, Air Force bombers dropped nearly three-quarters of all the satellite-aided GBU-31 joint direct attack munitions (JDAMs) that were delivered throughout the war. Air Force F-15E and F-16 fighters also contributed materially to strike operations after the tenth day, albeit in far smaller numbers, once the needed forward basing arrangements had been secured by flying long-duration

---

[1] For a full treatment of that joint air war, see Benjamin S. Lambeth, *Air Power Against Terror: America's Conduct of Operation Enduring Freedom,* Santa Monica, Calif.: RAND Corporation, MG-166-CENTAF, 2005.

sorties into Afghanistan from several friendly countries in the Persian Gulf. Nevertheless, carrier-based aviation operating from stations in the North Arabian Sea substituted almost entirely for what would have been a far larger complement of land-based fighter and attack aircraft in other circumstances because of an absence of suitable forward operating locations close enough to the war zone to make the large-scale use of the latter practicable.

The opening-night attacks were carried out by 25 Navy and Marine Corps F-14 and F/A-18 strike fighters launched from USS *Enterprise* and USS *Carl Vinson* operating in the North Arabian Sea, along with five Air Force B-1 and ten B-52 bombers operating out of Diego Garcia and two B-2 stealth bombers flying the entire distance from Whiteman AFB. These attack aircraft were supported by accompanying F-14 and F/A-18 fighter sweeps, as well as by radar and communications jamming provided by carrier-based EA-6B Prowlers.[2] In addition, a total of 50 Navy Tomahawk land-attack missiles (TLAMs) were also fired the first night against fixed high-priority targets by two Aegis destroyers, USS *McFaul* and USS *John Paul Jones*; a Spruance-class destroyer, USS *O'Brien*; and an Aegis cruiser, USS *Philippine Sea*, as well as two U.S. and British nuclear fast-attack submarines.

In a textbook example of integrated Air Force and Navy strike operations, the Navy's carrier-based fighters followed a designated ingress route over Pakistan (referred to informally by aircrews as "the boulevard") and checked in initially with an orbiting carrier-based E-2C Hawkeye, whose crew directed the flow of air traffic both inside Afghanistan and over the ocean approaches to it. Once the Air Force's longer-range E-3 AWACS arrived in the theater, it maintained contact with the CAOC in Saudi Arabia and assumed control of all air operations over Afghanistan, relieving the E-2C of that overland function because of the latter's substantially more limited endurance.[3] With the

---

[2] Thomas E. Ricks and Vernon Loeb, "Initial Aim Is Hitting Taliban Defenses," *Washington Post,* October 8, 2001.

[3] Robert Wall, "Navy Adapts Operations for Afghan War Hurdles," *Aviation Week and Space Technology,* November 19, 2001, p. 38. I had an opportunity to fly in that airspace myself and observe at first hand the flow of F/A-18 strike fighters from the carrier USS *John C. Stennis*

E-3 now on hand, E-2Cs continued to handle traffic departing from the carriers until incoming strikers reached the point over land where terminal control was handed over to the E-3. That function included controlling U.S. Air Force and British Royal Air Force (RAF) tankers as well, which established refueling tracks inside Afghan airspace toward the end of the war's major combat phase. Throughout that three-month phase, E-2Cs provided 24-hour air traffic management support, primarily by deconflicting the airspace over southern Pakistan.[4]

An elaborate joint Air Force and Navy inflight refueling scheme was employed, with carrier-based S-3 tankers orbiting off the coast of Pakistan to top off inbound Navy strikers just before the latter proceeded to their holding stations over Afghanistan. One carrier battle-group commander referred to these S-3 operations as a "bucket brigade."[5] Air Force KC-135 and KC-10 tankers, supplemented by RAF Tristars and VC-10s, orbited farther north to refuel the strikers again as their mission needs required before the latter returned to their ships.[6] These tankers flew thousands of miles from their forward operating locations to service the Navy's fighters and Air Force heavy bombers over Afghanistan. Strike missions from the carriers entailed distances to target of 600 nautical miles or more, with an average sortie length of more than four and a half hours and a minimum of two inflight refuelings each way to complete the mission.[7]

As indicated by statistics compiled by the CAOC during the 76 days of bombing between October 7, when Operation Enduring Freedom began, and December 23, when the major combat phase of the

---

into and out of Afghanistan during a 15-hour night E-3 AWACS combat mission out of Al Dhafra AB in the United Arab Emirates on April 20, 2007.

[4] Robert Wall, "Battle Management Dominates E-2C Combat Operations," *Aviation Week and Space Technology*, November 26, 2001, p. 40.

[5] Richard R. Burgess, "Air Strikes Hit Afghan Front Lines," *Sea Power*, December 2001, p. 25.

[6] Steve Vogel, "Gas Stations in the Sky Extend Fighters' Reach," *Washington Post*, November 1, 2001.

[7] Panel presentation on Operation Enduring Freedom by the participating carrier air wing commanders at the Tailhook Association's 2002 annual symposium, Reno, Nevada, September 6, 2002.

war ended after the collapse of the Taliban, some 6,500 strike sorties were flown by CENTCOM's forces altogether. Navy fighters contributed 4,900 of the strike sorties flown during that period, accounting for 75 percent of the total. For its part, although the Air Force flew only a quarter of the strike missions, its aircraft dropped 12,900 munitions, adding up to more than 70 percent of the total. The heavy B-52s and B-1s flew only 10 percent of the total strike missions, yet they delivered 11,500 of the 17,500 munitions, accounting for 65 percent of the total and 89 percent of all the munitions dropped by the Air Force.[8]

Much energy was wasted during the war's early aftermath in parochial fencing between some Air Force and Navy partisans over which service deserved credit for having done the heavier lifting in Enduring Freedom, with Air Force advocates pointing to the preponderance of munitions and overall tonnage dropped by the Air Force and Navy proponents countering that it was carrier-based aircraft, in the end, that flew the overwhelming majority of combat sorties and that performed nearly all of the "true" precision LGB attacks. To say the least, that fortunately fleeting contretemps was totally unhelpful to a proper understanding of what integrated Air Force and Navy strike operations actually did to produce such a quick and lopsided win over the Taliban and al Qaeda. True enough, two Navy carrier air wings were on station and ready for action long before the Air Force finally secured the needed forward basing to permit land-based fighter operations over Afghanistan. Even then, because of the relative remoteness of those bases, Air Force F-15Es and F-16s operating out of the Persian Gulf area flew only a small percentage of the overall number of fighter missions conducted in Enduring Freedom. That said, it bears noting that Air Force B-1 and B-2 bombers, with very few exceptions, dropped nothing but satellite-aided precision munitions of various types, and that Air Force B-52s dropped large numbers of accurate JDAMs in addition to unguided Mk 82 500-lb general-purpose bombs. It accordingly is a toss-up as to which service predominated in the precision-strike arena. Arguing over whether Air Force or Navy air power was more important in achieving

---

[8] William M. Arkin, "Old-Timers Prove Invaluable in Afghanistan Air Campaign," *Los Angeles Times*, February 10, 2002.

the successful outcome of Enduring Freedom was about on a par with arguing over which blade in a pair of scissors is more important in cutting the paper.

Finally, as many in the naval aviation community were among the first to acknowledge, without Air Force and RAF tankers to provide in-flight refueling support, the Navy's carrier air wings simply could not have conducted effective strike operations in other than the southernmost portions of Afghanistan.[9] As the CNO at the time, Admiral Clark, later put this point unabashedly: "I am very careful about making sure that my comments don't read, as some believe, that the Navy can do this by itself. It cannot. In fact, the early phase, the first several months of OEF, was totally and completely about the marriage and the union of the Air Force and the Navy. . . . We could have never, ever, conducted those missions without Air Force tanking—we couldn't even think about it. Air Force tankers made our success possible."[10]

Operation Enduring Freedom also featured the interplay of a veritable constellation of ISR systems that proved pivotal in enabling such tightly integrated Air Force and Navy strike operations. In a typical ISR fusion scenario involving carrier air assets, an Air Force electronic intelligence (ELINT) aircraft would note a spike in satellite phone traffic coming from a known Taliban location. An Air Force RQ-1 Predator unmanned aerial vehicle (UAV) would then be sent to the vicinity for a closer look, streaming real-time video of the building to targeteers in the CAOC in Saudi Arabia and at CENTCOM headquarters at MacDill AFB, Florida. A Navy F-14 pilot and RIO orbiting overhead

---

[9] This point was made emphatically by then-Rear Admiral James Zortman, who commanded the USS *John C. Stennis* battle group, during a panel discussion on carrier operations in Enduring Freedom at the Tailhook Association's 2002 annual symposium, Reno, Nevada, September 6, 2002.

[10] Marty Kauchak, "Navigating Changing Seas: Navy Chief Harbors No Illusions About the Challenges That Lie Ahead," *Armed Forces Journal International*, August 2002, p. 28. As a former naval aviator who commented on an earlier version of this report similarly observed: "In my day, for a Navy strike aircraft to go 600 miles was such an anomaly from both a distance and sortie-length perspective as to be not worth serious discussion. Now it's no big deal."

nearby would search for additional signs of activity and would visually confirm the suspected target to be a valid one. A combat operations officer in the CAOC, via the Air Force E-3 AWACS, would then read exact target coordinates to the F-14's RIO, who would finally program a JDAM for the assigned aimpoint on the building as soon as approval to drop was received. Navy F-14s also, for the first time during Enduring Freedom, showed their ability to derive and transmit accurate coordinates acquired by their LANTIRN targeting pod to inbound Air Force B-52s, whose crews, in turn, would use the information to drop JDAMs with great accuracy on approved targets. In addition, using a feature of their aircraft's onboard imagery capability called T3, for Tomcat tactical targeting, F-14 RIOs could determine a target's exact geographic coordinates and then pass those coordinates to the pilots of Air Force F-16s that lacked Global Positioning System (GPS) receivers or targeting pods and enable those Air Force pilots to drop GPS-aided cluster munitions.

In all, for the first time in the history of joint warfare, Operation Enduring Freedom showed real synergies in Air Force and Navy conduct of integrated strike operations. Navy fighters escorted Air Force bombers into Afghan airspace until allied air supremacy was established.[11] For its part, the Air Force (along with the RAF) provided roughly 80 percent of the tanker support (in aggregate terms of number of sorties flown and tons of fuel transferred) that allowed Navy carrier-based fighters to reach central and northern Afghanistan. That support, in turn, enabled sea-based strikes far beyond littoral limits,

---

[11] In light of the almost nonexistent Taliban air threat, it might seem at first look that these escort sorties were a complete waste of jet fuel and Navy assets. Yet that threat, although modest in the extreme, was not entirely inconsequential from a combat mission planner's perspective. The Taliban air arm included nearly 50 MiG-21 and Su-22 fighter aircraft, many out of service, that had been captured from defeated post-Soviet Afghan factions in 1996, and as many as 40 Taliban pilots were believed capable of getting those aircraft into the air. The fact that escort sorties were flown as a hedge against this barely credible threat again showed the extremely low risk tolerance of senior U.S. commanders and political leaders. In the circumstances, it was the Navy's carrier-based fighters operating in the North Arabian Sea that were uniquely able to meet this mission demand. (Lambeth, *Air Power Against Terror*, pp. 76–77.)

as well as a sustained carrier-based strike-fighter presence over remote target areas for hours if needed for on-call strikes on TSTs.

In addition, for the first time, naval aviators found themselves occupying key CAOC positions ranging from the deputy CFACC, then-Rear Admiral David Nichols, on down. These positions included the night CAOC director; the night guidance, apportionment, and targeting (GAT) cell director; and deputies for all key CAOC divisions (strategy, combat plans, combat operations, and ISR). The new arrangement also included carrier air-wing representatives and Navy platform and munitions experts as full-fledged members of the CAOC battle staff who were able to weigh in and effect operational changes as appropriate.

This unprecedentedly close harmony between the Air Force and Navy in joint operations over Afghanistan was said by both successive CFACCs, Air Force then-Lieutenant Generals Charles Wald and T. Michael Moseley, to have been outstanding at every level, from the commander of Task Force 50 and the Navy's representatives in the CAOC all the way down to the carrier air-wing commanders and aircrews who fought the war at the execution level.[12] For the first time ever, there was not the least apparent discord between Navy and Air Force personnel over the CAOC's operations. One informed report affirmed that the Navy was "fully committed to the [ATO] concept and [was] a full participant in the targeting process."[13] The first CAOC director for Enduring Freedom, Air Force then-Major General David Deptula, later added that throughout the campaign, there was "a coherent and cooperative group of planners from all the services, working together with a common goal and perspective" because they were all operating under one roof without barriers. "It just jelled," he said, in terms of personalities, adding that "we were all working together as an

---

[12] Conversations with then-Lieutenant General Charles F. Wald, USAF, Headquarters U.S. Air Force, Washington, D.C., May 15, 2002, and then-Lieutenant General T. Michael Moseley, USAF, Nellis AFB, Nevada, August 9, 2002.

[13] John G. Roos, "Turning Up the Heat: Taliban Became Firm Believers in Effects-Based Operations," *Armed Forces Journal International,* February 2002, p. 37.

air component, not as individual services."[14] The commander of the USS *Theodore Roosevelt* carrier battle group, then-Rear Admiral Mark Fitzgerald, likewise noted the "seamless interoperability" that he had experienced in working with Air Force aircrews and the CAOC. In marked contrast to the sometimes confused Desert Storm operations a decade before, carrier-launched strikes in Enduring Freedom benefited from totally integrated command and control, secure voice and digital data communications, secure videoteleconferencing, and real-time chat networks used by strike planners both afloat and in the CAOC ashore.[15]

These points were further affirmed by the combined force maritime component commander (CFMCC) in Enduring Freedom, Vice Admiral Charles Moore, Jr., who drew a sharp contrast between the spotty interservice command and control relationship that prevailed during the 1991 Persian Gulf War and that which characterized CENTCOM's air war over Afghanistan: "Joint interoperability was a significant issue in the past, but it's much less an issue today. I think it is constructive to compare what we are doing today with our experience during Operation Desert Storm. In that war, we had seams . . . in our displays of current operations. We had a couple of close calls [i.e., barely averted blue-on-blue encounters between friendly forces] because of that. Today, all of our commanders in the region can see the common operational picture right on their desktop computers. . . . All of our communications and most weapons are common, and our tactics, techniques, and procedures are all standardized. . . . When you look across the broad spectrum of warfare capabilities, we are almost totally interoperable with our joint and combined colleagues. During Enduring Freedom, we have reaped the benefit of the lessons learned from the past."[16]

---

[14] Quoted in Rebecca Grant, "The War Nobody Expected," *Air Force Magazine*, April 2002, p. 36.

[15] "USS *Theodore Roosevelt* Pounds Taliban and Terrorist Targets," *Sea Power*, December 2001, p. 24.

[16] Interview with Vice Admiral Charles W. Moore, Jr., USN, former Commander, U.S. Naval Forces Central Command and Commander, Fifth Fleet, "Committed to Victory," *Sea*

The uncommonly close meshing of carrier- and land-based air involvement in Operation Enduring Freedom, as well as the unprecedentedly prominent role played by the Navy in the CAOC throughout the war, bore witness to a remarkable transformation that had taken place during the years since Desert Storm, namely, a gradual convergence of Navy and Air Force thinking with respect to force employment at all levels of war. In previous years, "jointness" typically meant little more than the concurrent (and usually uncoordinated) participation of two or more services in a military operation. Yet as early as 1994, motivated in large part by the post–Cold War Navy needs identified by the 1991 Gulf War, then-Vice Admiral Owens, at the time the deputy CNO for resources, requirements, and assessments, introduced a new approach to Navy force planning aimed at increasing the service's leverage by seeking synergistic involvement with the Air Force when it came to expanded battlespace awareness, the military exploitation of space, and making the most of centralized command in joint air operations.[17] Two knowledgeable commentators on that history proved to have been more than a little prescient when they predicted, on the very eve of the September 11 attacks, that the coming year would witness "a triumph of the synergistic view of jointness . . . where the Navy and Air Force are concerned," with the result being the "closing of a promise-reality gap" that would yield "effects-based capa-

---

Power, March 2002, pp. 18–19. Admiral Moore went on to say (p. 20): "In past conflicts, right up through Kosovo, command and control of air operations has been a point of consternation within the joint force and the coalition, but it was not during Operation Enduring Freedom. I have to give the CFACC command a big 'Bravo Zulu' [Navy vernacular for 'well done'] for the terrific job they did. . . . We did not have a single complaint. We felt like our aircraft were being used exactly in the way they should be used, and that the CFACC fulfilled every single request we made of them. It was, I think, an extraordinary example of what we envisioned when we developed the doctrine of joint command and control of air operations."

[17] The first CAOC director for Enduring Freedom, Air Force Major General Deptula, summed it up this way: "What was amazing to me in the CAOC was the seamless nature of the way the components worked. . . . It was just great, it was so refreshing, particularly between the SOF [special operations forces] folks, the Navy, and us [in the Air Force]. . . . I have good things to say about the Navy. It really, really worked well in the CAOC." (Comments by General Deptula on an early draft of the author's RAND study *Air Power Against Terror*, January 24, 2004.)

bilities that are good for our regional commanders in chief and right for our nation."[18] That prediction was more than amply borne out by the experience of Operation Enduring Freedom that ensued beginning only a few weeks after.

---

[18] Barry and Blaker, "After the Storm," p. 130.

CHAPTER EIGHT

Further Convergence in Operation Iraqi Freedom

If Operation Enduring Freedom was tailor-made for integrated Air Force and Navy strike warfare, the subsequent three-week campaign in Iraq a year later to topple the Ba'athist regime of Saddam Hussein would prove to be no less so, particularly with respect to extended-range strike-fighter missions that were launched from the two Navy carriers that operated in the eastern Mediterranean Sea. In January 2003, in one of the first major deployment moves for the impending war, Secretary of Defense Donald Rumsfeld ordered the USS *Abraham Lincoln* carrier battle group to redeploy to the North Arabian Gulf from its holding area near Australia. The group was en route home from a six-month deployment in the Middle East but was directed to remain in CENTCOM's area of responsibility as a contingency measure. The *Theodore Roosevelt* battle group, just completing a predeployment workup in the Caribbean, was fresh to the fight and received orders to move as quickly as possible to reinforce *Constellation*, already in the Gulf, and *Harry S. Truman* in the eastern Mediterranean, for possible operations against Iraq. A fifth carrier battle group spearheaded by USS *Carl Vinson* moved into the Western Pacific to complement two dozen Air Force heavy bombers that had been forward-deployed to Guam. Air Force F-15Es were sent to Japan and Korea as backfills to cover Northeast Asia as USS *Kitty Hawk* moved from the Western Pacific to the North Arabian Gulf. In addition, the USS *Nimitz* battle group received deployment orders and got under way from San Diego in mid-January to wrap up an already compressed three-week training exercise, after which it headed for the Western Pacific. Finally, the

USS *George Washington* battle group, which had just returned to the East Coast in December following a six-month deployment in support of Operation Southern Watch, was placed on 96-hour standby alert, ready to return to Southwest Asia if required.[1]

By the end of the first week of March, the Navy had two carriers, *Theodore Roosevelt* and *Harry S. Truman*, on station in the eastern Mediterranean and three more carriers, *Kitty Hawk, Constellation*, and *Abraham Lincoln*, deployed in the North Arabian Gulf along with their embarked air wings, each of which included around 50 strike fighters.[2] In addition, *Nimitz* was en route to the Gulf to relieve *Abraham Lincoln*, which had been on deployment for an unprecedented nine months. The five carrier battle groups in position and ready for combat also included upward of 40 allied surface vessels and submarines armed with TLAMs. In addition, Air Force F-15Es, F-16s, and F-117s were in place at Al Udeid Air Base in Qatar, tankers and various ISR platforms were forward-deployed to Prince Sultan Air Base, Saudi Arabia, and more than 200 additional Air Force aircraft, including F-15s and F-16s, were positioned at two bases in Kuwait, with still more in Turkey, Oman, and the United Arab Emirates, all ready to carry out a multidirectional air attack. This fielded inventory of aircraft included 14 B-52s operating out of RAF Fairford in the United Kingdom and B-1 and B-2 bombers deployed to the Gulf region and Diego Garcia.[3] Four of the B-2s that would take part in the war were deployed from Whiteman AFB to Diego Garcia.[4]

Of the F/A-18 contingent that was committed to the impending campaign, 60 were Marine Corps Hornets attached to the 3rd Marine Aircraft Wing and operating out of land bases in the region in anticipated support of Lieutenant General James Conway's 1st Marine Expe-

---

[1] Scott C. Truver, "The U.S. Navy in Review," *Proceedings*, May 2003, p. 94.

[2] Robert Burns, "U.S. Gulf Force Nears 300,000 as Commander, Bush Consult," *Philadelphia Inquirer*, March 5, 2003.

[3] Eric Schmitt, "Pentagon Ready to Strike Iraq Within Days if Bush Gives the Word, Officials Say," *New York Times,* March 6, 2003.

[4] David E. Sanger with Warren Hoge, "U.S. May Abandon UN Vote on Iraq, Powell Testifies," *New York Times*, March 14, 2003.

ditionary Force.[5] In all, more than 700 Navy and Marine Corps aircraft figured in the total of 1,800 allied aircraft altogether that were committed to the major combat phase of Iraqi Freedom. That number included 236 Navy and carrier-based Marine Corps F/A-18s, 56 F-14s, 35 EA-6Bs, 40 S-3s, and 20 E-2Cs. The Marine Corps also provided another 130 land-based fighters and 22 KC-130 tankers. Those combined assets contributed to a coalition total of 1,801 aircraft, 863 of which were provided by the Air Force.[6]

A late-breaking development that threatened to impair severely the effective conduct of impending air operations, especially by the two Navy carrier air wings on station in the eastern Mediterranean, was Turkey's eleventh-hour denial of the use of its airspace by coalition forces. That denial promised to complicate the impending war effort greatly, since the carrier-based aircrews in the eastern Mediterranean had planned to transit Turkish airspace en route to targets in northern Iraq, with Air Force tankers supporting them out of Turkey's Incirlik Air Base. The Navy also had planned to fire TLAMs through Turkish airspace into Iraq. Without access to that airspace, one alternative would have been to reroute the carrier-based strike aircraft and TLAMs into Iraq from the west through Israeli and Jordanian airspace, which also would have made for an even shorter route than transiting Turkey's airspace. Some Bush administration officials pressed hard for the use of Israeli airspace if Turkey continued to balk on the issue. Secretary Rumsfeld, however, backed by his senior military advisers both in the Pentagon and at CENTCOM, concluded that attacking along a course that crossed the Jewish state would be too politically risky.[7] Alternatively, were Turkey's denial of needed access to persist, the two carriers in the eastern Mediterranean could redeploy through the Suez Canal to the Red Sea, with the result that they would be forced to

---

[5] Sanger, "U.S. May Abandon UN Vote on Iraq, Powell Testifies."

[6] Lieutenant General T. Michael Moseley, USAF, *Operation Iraqi Freedom—By the Numbers*, Shaw AFB, S.C.: Assessment and Analysis Division, Headquarters U.S. Central Command Air Forces, April 30, 2003, pp. 6–10.

[7] Thom Shanker and Eric Schmitt, "Rumsfeld Seeks Consensus Through Jousting," *New York Times*, March 19, 2003.

launch aircraft several hundred miles farther away from Iraq. The land-based tankers slated to support them would have to operate from even more distant shore bases.

As matters turned out, the aircraft of CVW-3 embarked in *Harry S. Truman* and those of CVW-8 in *Theodore Roosevelt* could not participate in strike operations against Iraq for the first two days of the war because they lacked permission to transit the airspace of any of the countries between the carrier operating areas and their likely targets in Iraq. Once Turkish airspace was made available to the coalition by D+3, however, numerous carrier-based strike sorties were finally flown over Turkey and, along with allied SOF teams on the ground, contributed to the early surrender of Iraqi army units fielded in the north.[8]

During the campaign's unpreplanned first night, thanks to what President George W. Bush and his principal deputies believed at the time to have been solid last-minute intelligence reporting that Hussein and his two sons were meeting at a known location in the Baghdad suburbs, Navy EA-6Bs provided electronic jamming support for Air Force F-15Es and RAF Tornado GR4s in opening a penetration corridor for the two Air Force F-117 stealth attack aircraft that led the ultimately unsuccessful decapitation attempt, followed shortly thereafter by 40 Navy TLAMS that were fired against the suspected meeting site. As before during Operations Allied Force and Enduring Freedom, the availability of Navy EA-6B jamming support was an iron-clad go/no-go criterion for all Iraqi Freedom strike missions, including those that involved Air Force B-2s and F-117s.

Later the next morning, when the Iraqis fired several Ababil 100 theater ballistic missiles at Kuwait in a response to the opening U.S. attack, the Navy's *Arleigh Burke*-class destroyer USS *Higgins* on station in the North Arabian Gulf served as a tactical ballistic missile early-warning system when her watch officers detected the incoming missiles within two seconds of their launch and cued Army Patriot surface-to-

---

[8] Philip P. Pan, "Turkish Leader Makes Request on Airspace," *Washington Post*, March 20, 2003.

air missiles based in Kuwait to engage and down them.[9] In yet another Air Force–Navy integrated strike operation, the ship's crew also transmitted launch-point information to the CAOC, which in turn targeted two Air Force F-16s that geolocated and destroyed the Iraqi missile launchers. In addition, as before during Operation Enduring Freedom, Navy carrier-based E-2Cs performed as airborne command and control centers in managing kill-box interdiction and CAS and helped direct Air Force fighters to their assigned target areas in Iraq. They also served as communications links between Air Force fighters, the CAOC, Air Force terminal attack controllers on the ground, and Army unit commanders; passed retasking orders and in-flight mission reports; and managed in-flight refueling by Air Force strike aircraft.

Similarly, Air Force B-1 bombers used their onboard moving-target indicator (MTI) radar in an ISR role to geolocate TSTs and transmit their coordinates to Navy strikers. Navy EA-6Bs worked with Air Force EC-130 Commando Solo and Compass Call aircraft in transmitting psychological warfare radio broadcasts and conducting radio network intrusions. The Air Force E-3 AWACS was used as a dynamic retasking tool to direct and redirect Navy strike aircraft. The Air Force E-8 joint surveillance target attack radar system (JSTARS) aircraft did the same during a three-day sandstorm that occurred during the first week of the campaign, as did a pair of Air Force RC-135 Rivet Joint aircraft, when Navy satellite-aided JDAMs were needed to replace LGBs that would not function to their fullest potential during the sandstorm. Once the sandstorm abated, Air Force RQ-1 Predator UAVs provided accurate target geolocation for Navy JDAM strikes. Air Force Special Operations Command joint terminal attack controllers on the ground also provided updated target coordinates for Navy JDAM attacks. And the single Air Force high-altitude RQ-4 Global Hawk UAV used its synthetic aperture radar (SAR) to detect an Iraqi missile that was partly hidden under a bridge and data-linked the missile's geographic

---

[9] Although this particular function did not entail a "strike warfare" activity, strictly speaking, it did enable a kinetic operation that occurred within the CAOC's field of regard and, as such, bears noting as a casebook example of seamless joint-force integration at the tactical level. It also was suggestive of comparable potential synergies between surface naval and land-based air assets that may now be ripe for further exploration and exploitation.

coordinates to a Navy F/A-18, whose pilot then attacked and destroyed the target by skipping a JDAM into it without damaging the bridge.

During the sandstorm that featured sustained winds of 25 knots gusting to 50 and visibility that was often reduced to less than 300 feet, the daily sortie rate of Air Force and Navy strikers did not substantially diminish. As Iraqi Republican Guard forces attempted to move under what their commanders wrongly presumed would be the protective cover of the sandstorm, aircrews from both services dropped satellite-aided JDAMs through the weather and destroyed hundreds of armored vehicles that were ferrying troops of the Medina Division toward forward elements of the U.S. Army's 3rd Infantry Division encamped about 50 miles south of Baghdad.[10] Such attacks by Air Force and Navy aircraft, both day and night and irrespective of weather conditions, were pivotal in reducing Iraq's conventional ground forces in the regular army and Republican Guard to a state of complete ineffectiveness as the three-week campaign entered its endgame.

Out of a total of 41,404 coalition sorties flown altogether during the major combat phase of Operation Iraqi Freedom, Navy and Marine Corps aircraft operating from carriers and large-deck amphibious ships flew nearly 14,000. Of those, 5,568 were fighter sorties, 2,058 were tanker sorties, 442 were E-2C sorties, and 357 were ISR sorties. In addition, more than 800 TLAMs were launched from various other ships assigned to the five carrier battle groups, and Navy assets flew 25 percent of the theaterwide ISR sorties during the three-week period of major combat.[11]

Operation Iraqi Freedom also set a new record for close Navy involvement in the high-level planning and conduct of joint air operations. As the deputy CFACC once again for Iraqi Freedom, then-Rear Admiral Nichols (who had previously commanded NSAWC at Fallon) was not just the "senior naval representative" in the CAOC but General Moseley's alter ego, to all intents and purposes, when it came to

---

[10] Peter Baker and Rajiv Chandrasekaran, "Republican Guard Units Move South from Baghdad, Hit by U.S. Forces," *Washington Post*, March 27, 2003.

[11] Vice Admiral Michael Malone, USN, "They Made a Difference," *The Hook*, Summer 2003, p. 26.

commanding and managing the air war. In addition, alternating with Air Force Colonel Doug Erlenbusch, Navy Captain Russell Penniman was co-director of the combat plans division in the CAOC, which did all of the target analysis and weaponeering.[12] Navy Captain (now-Rear Admiral) William Gortney was the naval air liaison coordinator in the CAOC. That representation and more stood in marked contrast to the Navy's less gratifying experience 12 years before during Operation Desert Storm, when the overwhelming majority of the staffing of the CAOC's targeting cell was by Air Force officers, with Navy members both too few in number and too junior in rank to wield significant clout in influencing the day-to-day decisionmaking.[13]

According to then-Vice Admiral Timothy Keating, who was dual-hatted as Fifth Fleet commander and CENTCOM's CFMCC for Iraqi Freedom, detailed and coordinated planning had taken place beforehand during the buildup for the campaign between Fifth Fleet and U.S. Central Command Air Forces (CENTAF) in determining which personnel the Navy would send to the CAOC and what their qualifications needed to be. If a qualification requirement was not met, selected individuals received rush schooling in the needed skills to enable them to augment the CAOC staff. Admiral Nichols had an augmentation team of 101 Navy personnel, around half of whom were reservists and the rest of whom were drawn from the Navy's aviation weapons schools, NSAWC, and fleet units. That made for an unprecedentedly visible and influential Navy presence in key CAOC leadership positions. In the end, the Navy filled 20 percent of the CAOC's manning billets. The naval reservists assigned to the CAOC had the added advantage

---

[12] Kim Murphy and Alan C. Miller, "The Team That Picks the Targets," *Los Angeles Times*, March 25, 2003.

[13] For more on that earlier experience from the perspective of a participating naval officer, see Commander Daniel J. Muir, USN, "A View from the Black Hole," *Proceedings*, October 1991, p. 86. It should be further noted that the largely junior-officer Navy representatives in the CAOC's targeting cell during Desert Storm had only had previous hands-on exposure to single air-wing, Navy-only air operations and lacked significant familiarity with the more complex and demanding requirements of planning and conducting integrated joint and combined air operations.

of coming from the Second and Third Fleet staffs and of being both JFACC-trained and well versed in the processes of the CAOC.

Furthermore, at the operational and tactical levels, the five carrier air wings that took part in the campaign were better integrated into the ATO process than ever before, with each air wing having designated representatives in the CAOC where the daily ATO was assembled to ensure that the wings were assigned appropriate missions. The wings also had ready access to onboard software in their carriers' strike planning spaces that automatically searched the complex ATO for Navy-pertinent sections, eliminating any need for air-wing mission planners and aircrews to study the entire document. Closer cooperation in recent years between the Air Force's and Navy's air warfare centers at Nellis and Fallon, respectively, yielded major dividends in improved joint-force interoperability. The Air Force also routinely shared munitions with the Navy as necessary to ensure the fullest possible effectiveness of both services. The Air Force chief of staff during Iraqi Freedom, General John Jumper, later reported that because LGBs suffered reduced effectiveness during the sandstorm, the Air Force provided the Navy with the JDAMs it needed for its air wings to continue meeting their daily ATO tasking and thereby keep the pressure on.[14]

In sum, Operation Iraqi Freedom was a true joint-service effort involving wholly integrated Air Force and Navy strike operations. In the apt words of two historians writing an early synopsis of the war, that effort saw "little of the petty parochialism that too often marks interservice relations within the [Washington] Beltway."[15] The chairman of the Joint Chiefs of Staff, Air Force General Richard Myers, later observed that the close integration not just of Air Force and naval strike assets but of *all* allied force elements was "a huge lesson here."[16] Speaking as the CFMCC for Iraqi Freedom, Admiral Keating characterized the

---

[14] "Air Force Chief Says Munition Stockpiles Are Sound," *Inside the Air Force*, April 11, 2003, p. 11.

[15] Williamson Murray and Major General Robert H. Scales, Jr., USA (Ret.), *The Iraq War: A Military History*, Cambridge, Mass.: The Belknap Press of Harvard University Press, 2003, p. 114.

[16] Vince Crawley, "Less Is More," *Army Times*, April 21, 2003.

operational payoff of all this as "joint warfighting at the highest form of the art I'd ever seen. . . . There was understanding, friendship, familiarity, and trust among all the services and special forces working for [Army] General [Tommy] Franks [the overall joint-force commander for the three-week campaign]. He did, in my view, a remarkable job of engendering that friendship, camaraderie, and trust. In fact, he insisted on it. . . . There was no service equity infighting—zero."[17]

---

[17] Interview with Vice Admiral Timothy J. Keating, USN, "This Was a Different War," *Proceedings*, June 2003, p. 30.

CHAPTER NINE

Emergent Trends in Air Force–Navy Integration

The performance of Air Force and Navy strike assets in the first two American wars of the 21st century bore ample witness to the giant strides that have been made in the integration of the two services' air warfare repertoires since Operation Desert Storm. The two wars saw naval aviation more fully represented than ever before throughout the CAOC. They also saw it fully integrated into the joint and combined air operations that largely enabled the successful outcomes in each case. Finally, both wars showed increased Air Force and Navy acceptance of effects-based thinking and planning, as well as a common use by the two services of the joint mission planning system that more recently has been reflected in Air Force participation in joint planning for contingency operations in the Seventh Fleet's area of responsibility in the Western Pacific. As attested by the Navy's experience in both Enduring Freedom and Iraqi Freedom, the CAOC-generated ATO is now disseminated electronically to carrier strike groups in an easily usable and manipulatable form and is updated hourly for each carrier via SIPRNet email.[1] Moreover, prompted by the experience of Enduring Freedom and Iraqi Freedom, prospective carrier air-wing commanders and other naval aviation leaders now routinely spend upward of 100 days forward-deployed in CENTAF's new CAOC at Al Udeid Air Base in Qatar for operational planning familiarization in a senior CAOC staff assignment before assuming their new command responsibilities. They also routinely attend the Air Force's strike planning course at Hurlburt

---

[1] SIPRNet is an abbreviation for "secure internet protocol router network."

Field, Florida, and, after having moved on to post-command billets, its week-long CFACC course at Maxwell AFB, Alabama.

In addition, as indicated by the many examples arrayed above, major progress has been progressively made in recent years in the realm of increased Air Force and Navy mutual understanding and shared beliefs with respect to the planning and conduct of joint strike warfare. Even before the most recent tests of Operations Enduring Freedom and Iraqi Freedom, the ever-improving performance of U.S. carrier air power in repeated post–Cold War contingency operations showed that although, as two informed observers noted, the Navy and Air Force continued "to stage knife fights over doctrine," still had problems communicating with one another during operations, and continued "to squabble—quietly or not—over their 'fair shares' of the defense budget," the two services had in fact become much better at operating in a joint setting than had been the case a decade before.[2] Thanks to that steady convergence, most of the former parochial tensions between the two services, such as the Navy's long-standing difficulty with the Air Force–inspired joint air tasking process discussed above, are now gone at the operational and tactical levels of war.

Some of this progress dates as far back as the early aftermath of Desert Storm. For example, in response to the problem of getting the ATO electronically transmitted to the carrier air wings aboard ship that the Navy identified during the first Gulf War, the office of the CNO announced a decision in 1992 to equip both numbered fleet and Middle East Force flagships and all further deploying aircraft carriers with the Air Force's CAFMS, thereby allowing strike planners afloat to nominate aircraft and TLAM missions and targets to the JFACC, as well as easing their timely receipt of the daily ATO. The intent of this move was to allow the Navy to work more closely with the Air Force toward fielding a common system for integrating such ATO information as launch time, targets, and required weapons with the naval intelligence processing system afloat, thereby providing an interactive link

---

[2] Barry and Blaker, "After the Storm," p. 117.

with the air component commander that would allow due Navy input into the air tasking process from initial planning through execution.[3]

Moreover, it was not just a matter of the Navy's accommodating the Air Force by seeking ways to work more easily with the latter's ingrained practices. At about the time that CENTCOM was gearing up for the major combat phase of Iraqi Freedom, the Air Force chief of staff at the time, General Jumper, frankly conceded, if implicitly, that some Navy critics of the inefficiencies of the ATO process had a legitimate basis for their discontent. Said General Jumper: "We take a rap in the Air Force about having a 72-hour ATO cycle. . . . It is really not true. It's the planning cycle that is 72 hours. The execution cycle can be instantaneous." However, he went on to note, "there is a point to that argument. . . . You go into an AOC today, and what will you see? Tribal representatives sitting down in front of tribal workstations, interpreting tribal hieroglyphics to the rest of us who are on watch. And then what happens? They stand up and walk over to another tribal representative, and reveal their hieroglyphics, which are translated by the other tribe into its own hieroglyphics and entered into its own workstation."[4] His point was a need for tighter horizontal integration of command and control both within each service and across service lines in the interest of eliminating such inefficiencies and shortening the sensor-to-shooter connection.

Viewed in hindsight, the Air Force and Navy entered the information revolution mostly along parallel but separate paths rather than jointly. However, with the advent of the global command and control system, Link 16, and related connectivity improvements, the prospect has finally emerged of joint operations by the two services that entail what two early commentators on air-naval integration called "true interoperability, functional integration, and order-of-magnitude improvement in capability."[5] This welcome prospect emerged in part out of the Navy's development of its cooperative engagement capabil-

---

[3] John F. Morton, "The U.S. Navy in 1991," *Proceedings*, Naval Review 1992, p. 136.

[4] Mark Hewish, "Out of CAOCs Comes Order," *Jane's International Defense Review*, May 2003, p. 22.

[5] Barry and Blaker, "After the Storm," p. 118.

ity (CEC) during the waning years of the Cold War. With the latter's stress on space-based surveillance and the need to develop and be capable of reacting to a common operating picture, CEC laid down the initial needed groundwork for closer operational convergence of the Navy with its Air Force sister service. As early as 1993, the Navy initiated a demonstration of this capability and its potential by linking the commander in chief of U.S. Atlantic Fleet with the Air Force's Air Combat Command and Army Forces Command, with a subsequent inclusion of Fleet Marine Forces Atlantic.

In addition to its primary goal of providing expanded maritime domain awareness, the Navy's CEC also offered the potential for expanded battlespace awareness and a common operating picture of littoral force dispositions. As a buy-in cost for this, however, it required a more centralized command and control arrangement than the Navy was accustomed to in order to work most efficiently. Thus were planted the seeds of a growing convergence by the Air Force with the Navy's concept of network-centric warfare. As the commander of Naval Air Systems Command put it in 1999, "we have spent this whole decade concentrating on better interoperability. We learned a lesson in Desert Storm that we have to pay more attention to operating with our counterparts. . . . We must be able to communicate freely—both in planning and in operations—and many of the systems we have in development or deployed today are aimed specifically at improving that ability."[6]

By way of some essential background to this development, the Navy's newfound littoral focus that emanated from its experience in Desert Storm led to an effort to leverage new technology to project naval power ashore by dominating not just the sea but a five-dimensional battlespace that also included the air, land, space, and cyberspace. An associated idea was for the Navy to become an enabling force configured and poised to lay the groundwork for the subsequent introduction of Air Force and Army force elements into the war zone. The overarching goal of that vision was allowing for deep attack from the

---

[6] Interview with Vice Admiral John A. Lockard, USN, "Affordable Readiness for the Operator: 21st-Century Aviation Solutions Enable Dominance from the Sea," *Sea Power*, June 1999, p. 8.

sea by a studied shift from platform-centric to network-centric operations and by concentrating on what one retired admiral called "force
posture rather than force structure—the way the Navy *operates* rather
than the way it looks."[7]

Network-centric warfare is not, of course, unique to the nation's
sea-going forces. Rather, it entails an approach to precise and agile
force employment in all mediums that seeks to move beyond platform-
centric operations and reliance on stovepiped and single-purpose systems and to leverage the increasingly linked and unifying meshwork
of the entire American military information grid. In essence, network-
centric operations are information-intensive interactions between the
many computational nodes on the network. The idea behind them was
first articulated in broad outline by Vice Admiral Arthur Cebrowski
in 1998 in an amplification on CNO Admiral Jay Johnson's earlier
observation that the Navy, like the American defense establishment as
a whole, was in the process of undergoing "a fundamental shift from
what we call platform-centric warfare to something we call network-
centric warfare."[8]

The idea of the latter drew its inspiration from an obscure but
powerful notion called Metcalfe's Law propounded by the creator of
the Ethernet, Robert Metcalfe, which stipulates that the robustness
and capability of a network is directly proportional to the square of the
number of nodes in the network. As Admiral Cebrowski explained it,
the value of a network "increases as information moves toward 100-
percent relevant content, 100-percent accuracy, and zero timeline
delay—toward information superiority." The sort of network-centric
operations enabled by this capability, he added, promise to produce a
shift from attrition-based warfare to "a much faster and more effective
warfighting style characterized by new concepts of speed of command

---

[7] Vice Admiral Scott Redd, USN (Ret.), "Sustained Assured Access: The Navy's Role in Joint
Warfare," *Armed Forces Journal International*, September 2001, p. 68, emphasis added.

[8] Vice Admiral Arthur K. Cebrowski, USN, and John J. Garstka, "Network-Centric Warfare: Its Origin and Future," *Proceedings,* January 1998, p. 29.

and self-synchronization."[9] An example might be a SEAD attempt featuring not just individual air attacks against individual SAM sites to shut down radars and temporarily suppress them but additional shooters as well, such as land- or carrier-based strike fighters and possibly also ATACMS (Army tactical missile system) fires targeting an entire engagement grid, such that all sites would be negated or destroyed simultaneously.

As Admiral Fitzgerald said later of the Navy's early-generation network-centric applications, the carrier air contribution to Operation Enduring Freedom revealed "the tip of the revolution that's continuing" in naval aviation. The first revolution was in the realm of precision weapons and precision targeting. The second, the admiral said, will entail "going from the analog to the digital age in communications architecture." During Desert Storm in 1991, email was not available at all aboard Navy ships, available bandwidth was measured in bits, and, as noted earlier, the ATO had to be delivered to the carriers in hard copy in daily runs by courier flights. During Operations Enduring Freedom and Iraqi Freedom, in sharp contrast, the ATO was updated for each participating air wing almost hourly via email. Similarly, during Desert Storm, almost every strike-fighter target was extensively prebriefed. A dozen years later, carrier air wings were routinely launching aircraft from carrier decks either without assigned targets or with preassigned target packages that were changed in midmission based on fast-breaking updates in near-real time. During its first war against Iraq, the Navy communicated in an almost totally analog fashion over the radio. By the time of Operations Enduring Freedom and Iraqi Freedom, however, the Navy, like the Air Force, was at the brink of a completely digital age. Thanks to a multitude of innovations in digital communications and linked architectures, the conduct of command and control in a carrier battle group was now done mostly through email, with information also being passed digi-

---

[9] Cebrowski and Garstka, "Network-Centric Warfare," p. 31.

tally back and forth between airborne Air Force and Navy aircraft and terrestrial stations both ashore and afloat.[10]

Indeed, the second Gulf War in 2003 featured a more closely linked U.S. force than ever before. As one CENTCOM staffer put it: "Everything that had a sensor was connected."[11] To note a representative example, the aircraft carrier USS *Abraham Lincoln* featured a joint fires network and CEC system that allowed strike-group participants to share radar information and to fire missiles based on off-board information provided by other ships in the battle group. This capability was expanded with the arrival of the carrier USS *Nimitz* and the first Navy E-2Cs equipped with the system. The joint fires network, a Navy adaptation of the Army's tactical exploitation system (TES) sensor fusion mechanism, allowed carriers to receive imagery from airborne platforms and signals intelligence (SIGINT) from the Air Force's RC-135 Rivet Joint. Similarly, the multifunction information distribution system (MIDS), a nodeless and secure Link 16–based jam-resistant tactical data link, also made a major difference in enabling enhanced interoperability with other joint and multinational platforms equipped with that capability. Now in the fleet and with more than a thousand Link 16 terminals in the four services, it was a major contributor toward the continuing transition from analog to digital warfighting, and it paved the way for the next step in network-enabled operations.[12] As Admiral Fitzgerald suggested, "where we are trying to get to ultimately is . . . a fully linked [joint] force. We want to have 'Internet-in-the-cockpit' capability. We want to have the ability for a pilot flying in harm's way to call up and say, 'OK, what is my target looking like right now? What is the latest that I am getting out of that Predator or that Global Hawk right now?" He added: "Transformation is no longer in the platform.

---

[10] "Fitzgerald: Recapitalization Poses Challenge for Naval Air," *Sea Power*, March 2004, p. 28.

[11] Thomas E. Ricks, "What Counted: People, Plan, Inept Enemy," *Washington Post*, April 10, 2003.

[12] Captain David C. Hardesty, USN, "Fix Net Centric for the Operators," *Proceedings*, September 2003, p 69.

The platforms we are fighting with today are the platforms that we fought with during Desert Storm. . . . What [has] changed are the sensors, the weapons, and the [data] links."[13]

In a major move toward acting further on this emerging cross-service potential, officials from the Air Force's Command and Control (C2) Constellation and the Navy's FORCENet initiatives joined forces shortly after the major combat phase of Iraqi Freedom ended to work toward developing a common architecture for network-enabled operations (another, and perhaps more useful, way of saying "network-centric warfare"), with a view toward merging their still largely service-specific nets into something that might eventually be used by all four services. (The Air Force's C2 Constellation network connects such Air Force sensor platforms as Global Hawk, AWACS, and Predator UAVs. Its ultimate aim is to provide uniform information to all activities in the kill chain. For its part, FORCENet is the overarching term used by the Navy to describe its ongoing process of applying network-centric theory to create a similar seamless grid of internetted sensors, weapons, individuals, and command and control mechanisms that are accessible to all elements of the fleet and are intended to enhance their ability to sense, locate, communicate, attack, and assess.) Similar cross-service efforts were initiated in such related areas as ISR management, target aimpoint generation, tactical data links, and joint tactical radios. This accelerated move to lash up the respective network-enabling capabilities of the two services has been rendered substantially easier than it would have been otherwise because Air Force and Navy strike-warfare operations have, at long last, become so closely comparable in both planning and execution. Navy FORCENet officials have consequently viewed the strike-warfare arena as a lucrative one in which to test the ability of the two services to converge on network-centric concepts and baseline capabilities, which might then gravitate to U.S. Joint Forces

---

[13] Hunter Keeter, "Navy's Lessons from Afghanistan, Iraq Include Networked Tactical Aircraft," *Defense Daily*, May 6, 2003, p. 3.

Command (JFCOM) as the natural entity to oversee such a joint enterprise.[14]

In early 2004, the Air Force and Navy took the important next step of actually drafting joint documents that would chart a course for making their respective network-centric infrastructures and concepts of operations more interoperable. That collaborative move grew out of informal conversations between then-Lieutenant General Thomas Hobbins, at the time the director of Air Force warfighting integration, and then-Vice Admiral John Nathman, at the time the deputy CNO for warfare requirements and programs. That dialogue eventually led to the establishment of a multiservice command, control, communications, and computers (C4)/ISR integration working group to try to merge the respective plans of the two services in this broad area. As explained by the chief of the C2 Constellation division of the Air Force Command and Control and Intelligence, Surveillance, and Reconnaissance Center at Langley AFB, Virginia, Lieutenant Colonel Rick Painter, "we're trying to talk early about those architecture views so that they are more comprehensively joined when they get completed."[15]

Most of the parallel Air Force and Navy programs of concern in this respect are connected to common airborne missions between those two services. As an example of the kinds of better integration that is being sought by this effort, the two services were looking at a common upgrade of their separate target mensuration systems, with the Air Force using a system called Raindrop and the Navy using a different system called precision-targeting workstation. Both were aiming to acquire a common mensuration tool. A related initiative entailed a slowing down in late 2003, ahead of a contract being issued, of an Air Force–led program to upgrade the distributed common ground system (DCGS) that receives and processes intelligence gathered by airborne platforms so that the initiative could be responsive to Navy operational

---

[14] John T. Bennett, "USAF Constellation, Navy FORCENet Step Up Interoperability Efforts," *Inside the Pentagon*, July 31, 2003, pp. 1, 16.

[15] Hampton Stephens, "Air Force, Sister Services Begin Work to Link Network Warfare Visions," *Inside the Air Force*, February 27, 2004, p. 12.

requirements. Still another entails a search for greater commonality between Air Force and Navy command and control systems. By early 2004, the two services had completed a draft joint policy document to guide technology integration toward the development of a common sensor strategy. The avowed goal of both is to use the Air Force's C2 Constellation and the Navy's FORCENet in a way aimed at linking the entire gamut of airborne sensors that each service uses to collect operational-level intelligence to further shorten the time required to find, fix, target, track, engage, and assess a TST.

As for other signs of progress toward greater cross-service integration in strike warfare, there have been steady improvements in joint operations and training between the Air Force and Navy since American combat involvement in Vietnam ended more than three decades ago. For years, naval aviators have routinely taken part in the Air Force's recurrent Red Flag realistic large-force employment training exercise that first began in late 1975 and that continues to be conducted roughly six times a year within the instrumented Nellis range complex. Also, the Air Force's and Navy's undergraduate pilot training (UPT) programs are now fully integrated, with Air Force officers commanding Navy primary UPT squadrons and vice versa. The two services continue as well to provide exchange officers to each other's line squadrons and flight test units on a regular basis, with a Navy lieutenant commander recently assigned to fly the F-22A Raptor fifth-generation Air Force fighter with the 422nd Test and Evaluation Squadron at Nellis. In addition, Navy E-2C Hawkeye crewmembers regularly fly aboard the Air Force's E-3 AWACS whenever there is an operational need for their presence at the console. Similarly, ever since the Air Force retired its EF-111 electronic warfare aircraft from service not long after Desert Storm, Air Force aircrews have routinely been assigned to full tours of duty as serving aircrew members with the Navy's EA-6B shore-based expeditionary squadrons.

Furthermore, there has been recurrent cross-communication and cross-fertilization between the Air Force's and Navy's weapons schools in an instructor exchange program that has experienced ups and downs since its inception in the late 1970s. During the late 1980s and early 1990s, when the Air Force Weapons School produced three classes per year, the instructor exchange was a standard twice-yearly exercise. Navy TOPGUN instructors from Miramar would fly to Nellis for a week in April and Air Force Weapons School F-15Cs would, in turn, fly to Miramar for a week in August, with each deployment offering unmatched opportunities for both sides to exchange instructional techniques and procedures, as well as to compare tactics, syllabi, aircraft capabilities, and the like.

When the Air Force Weapons School went to two classes a year and the Navy's TOPGUN program moved from Miramar to Fallon, those initial exchanges began to die on the vine, with the last one occurring in 1999 until the most recent commander of the Air Force Weapons School's F-15C Division pressed hard to reestablish the program, with the first renewed exchange taking place in June 2006. The exchange was adjudged by all participants to have been a great success, with useful and important lessons learned by both sides through focused discussion and comparison of such areas as

- current tactics
- course syllabi
- threat replication
- communications standards
- new capabilities in the F-15 and F/A-18
- briefing, execution, and debriefing standards
- facilities and support assets
- current and future conflict issues.

In addition, Air Force and Navy weapons-school instructors during the exchange week at Fallon flew daily offensive and defensive counterair missions that pitted integrated F-15 and F/A-18 attack packages against a mix of adversary formations consisting of F-5s, F-15s, F-16s, F/A-18s, and former Israeli Air Force Kfirs. In these exercises, Air Force Weapons School instructors planned, briefed, led, and debriefed the morning missions, with TOPGUN staff instructors leading the afternoon events in the same scenario. During the course of the exchange, it quickly became apparent that Air Force and Navy tactics had diverged since the last exchange six years previously, and both sides concluded at week's end that the cross-talk gained in post-mission debriefings had been exceptionally valuable all around. As a result, a repeat of the exchange was scheduled for June 2007, with the Air Force Weapons School's F-15C Division commander having arranged to host a TOPGUN deployment to Nellis to keep a resurgent yearly exchange program going. As he concluded from the 2006 experience, "there were great lessons on each side, something that you could never get without face-to-face flying and debriefing in secure facilities, which we were able to do. It was the best $50,000 we ever spent."[16]

In addition, Air Force and Navy fighter employment TTPs are now coordinated on a regular basis between Nellis and Fallon and, as a result, are completely common, as are the classified joint-service "three-dash-one" publications that promulgate agreed-on concepts of operations to be used against specific enemy threat systems. Last, to cite but one more of a whole raft of similar examples that could be listed, common terminology and brevity codes for radio voice communications are now standard for both services for CAS and air-to-air combat, and agreed and formalized joint CAS doctrine is being implemented in both services. (This last and most recent development was prompted by some disturbing instances of friction in timely CAS delivery to beleaguered U.S. ground forces that both services unexpectedly encountered during Operation Anaconda in Afghanistan in March 2002.)

Perhaps most constructively of all, the two services continue to bring their respective forces and combat-support assets together in a variety of joint training and experimentation exercises aimed at further honing their interoperability and extracting the most from their synergistic potential when it comes to the conduct of effective strike operations. One such recent exercise in which an instructive air-naval integration precedent was established brought Air Force and Navy air assets together in the vicinity of Alaska in a scenario that focused on homeland security and entailed military responses to a range of simulated natural disasters and terrorist events, including earthquakes, biological and chemical attacks, and terrorist events in the air and at sea.

---

[16] Email communication to the author from Lieutenant Colonel Andrew Croft, USAF, Commander, 433rd Weapons Squadron, Nellis AFB, Nevada, April 4, 2007.

That annual evolution, Exercise Northern Edge 2005, was conducted by U.S. Northern Command. It featured the involvement of both a Navy surface maritime action group and multiple Air Force, Navy, and Coast Guard aircraft that took part in various at-sea deterrence and defense operations during the five-day event from August 15 to August 19, 2005. Participating aircraft included an Air Force E-3 AWACS and a Coast Guard C-130, along with a Navy SH-60B Seahawk helicopter and P-3 Orion aircraft that conducted maritime search and surveillance operations. In addition, Air Force F-15E strike fighters performed low-pass show-of-force operations over notional suspect vehicles. Significantly during this exercise, the Navy, for the first time, exercised tactical control of an Air Force AWACS in a maritime-operations scenario, and the participating Air Force F-15Es were also controlled by the Navy from the flagship destroyer USS *Russell*. After the exercise ended, the commander of Destroyer Squadron 21 and the maritime action group commander, Navy Captain Vic Mercado, reported that "the coordinated joint surveillance resulting in the call for a show-of-force by the [Air Force] fighters was a highlight for the maritime operations, because it demonstrated a key exercise objective of cooperation and interoperability among the services for homeland defense."[17]

Most recently, such joint Air Force and Navy involvement in realistic large-force training in a maritime setting occurred during Exercise Valiant Shield '06, a five-day evolution conducted in the vicinity of Guam from June 19 to June 24, 2006, under the command of Admiral Gary Roughead, USN, the commander of U.S. Pacific Fleet, who served as joint-force commander for the exercise, with Air Force Lieutenant General David Deptula, commander of PACAF's Kenney Warfighting Headquarters at Hickam AFB, Hawaii, as his JFACC and with Rear Admiral Mark Emerson, commander of NSAWC at Fallon, assigned as deputy JFACC for the exercise. Valiant Shield involved the participation of some 22,000 personnel, 280 aircraft, and 30 ships,

---

[17] Lieutenant Denise Garcia-Barham, USN, "U.S. Warships Complete Exercise in Alaska," *Navy NewsStand*, August 26, 2005. I am grateful to Captain Mercado for having brought this experience to my attention during a roundtable discussion on air-naval integration issues aboard the aircraft carrier USS *John C. Stennis* at sea on April 26, 2006.

including the aircraft carriers USS *Kitty Hawk*, *Abraham Lincoln*, and *Ronald Reagan* and their three embarked air wings. It was the largest military exercise conducted in Pacific waters since the Vietnam War and represented the first installment of what will become a regular biennial exercise series involving various U.S. service branches and communities. It focused mainly on joint detection, tracking, and engagement of forces at sea, on land, and in the air in response to a broad spectrum of operational challenges.

After the exercise ended with nearly 2,000 sorties having been flown by all participating aircraft, General Deptula characterized it as "an opportunity to interface large numbers of [American] air and sea forces together in a unique environment and to work out some of what we call frictions. . . . You find out things that might not go as you would have anticipated or planned. These types of exercises allow us to work out those challenges in advance." On the operational synergy that was sought and achieved during the course of the joint-force exercise, he added: "We're not interested in what Navy or Air Force airplanes are doing separately. We take the approach that air power is air power, and we're interested in ensuring [that] we take a unified stance in working those assets together with our sea-based assets in achieving the commander's overall objectives."[18]

Over the course of the five-day exercise, which included the participation of the Navy command ship USS *Blue Ridge* in addition to the three carrier strike groups, Air Force B-2 stealth bombers assigned to the 509th Bomb Wing at Whiteman AFB, Missouri, were joined by F-15Cs from the 18th Wing home-stationed at Kadena AB, Okinawa; by F-16CJs from the 35th Fighter Wing at Misawa AB, Japan; by F-15Es from the 3rd Wing at Elmendorf AFB, Alaska; and by a host of additional Air Force tanker and airlift aircraft, as well as by six Marine Corps F/A-18Cs from VMFA-97 based at Iwakuni AB, Japan. While the exercise was under way, the commander of Carrier Strike Group 7 aboard USS *Ronald Reagan*, Rear Admiral Michael Miller, commented

---

[18] Captain Yvonne Levardi, USAF, "Air Ops Center Wraps Up Valiant Shield," news release, Office of Public Affairs, Kenney Warfighting Headquarters, Hickam AFB, Hawaii, June 26, 2006.

that "joint interoperability is the key to successfully responding to future contingencies in the Pacific. Exercises such as Valiant Shield give us an opportunity to ensure [that] joint command, control, and communication procedures are seamless."[19] Echoing that perspective, Air Force Major Paul Hahn of the Kenney Warfighting Headquarters' Combat Operations Division later concluded: "We had a very successful exercise. This was a great opportunity to practice joint interoperability with our Navy counterparts as if it were a real-world situation."[20]

---

[19] Shane Tuck, "Valiant Shield Provides Valuable Joint Training Among U.S. Military Forces, *Navy NewsStand*, June 20, 2006.

[20] Levardi, "Ops Center Wraps Up Valiant Shield."

CHAPTER TEN

A New Synergy of Land- and Sea-Based Strike Warfare

As described in broad outline above, the unprecedentedly close integration of Air Force and Navy aerial strike operations during the first two American wars of the 21st century handily confirmed the observation of a respected specialist in ship design and broader sea power issues when he wrote in 1998 that "carrier-based and land-based tactical aircraft, as well as the CONUS-based Air Force bomber force, are intertwined in their support of each other."[1] To be sure, the two services have long paid lip service to their mutually reinforcing potential in their declaratory rhetoric. Yet in the increasingly competitive annual budget battles within the Pentagon, the strike-warfare components of the Air Force and Navy have all too often appeared as though they were mainly devoted to putting each other out of business.

The real-world experience described above, however, strongly suggests that when it comes to the crucial matter of integrated strike-warfare operations, the two services should consider one another natural allies in the roles and resources arena, since they did not compete with each other in Operations Enduring Freedom and Iraqi Freedom but rather mutually supported and reinforced one another in the successful pursuit of joint campaign objectives. Indeed, when viewed from an operational rather than a bureaucratic perspective, the Air Force's and Navy's long-standing involvement in air-delivered conventional

---

[1] Reuven Leopold, *Sea-Based Aviation and the Next U.S. Aircraft Carrier Design: The CVX*, MIT Security Studies Program Occasional Paper, Cambridge, Mass.: Center for International Studies, Massachusetts Institute of Technology, January 1998, p. 11.

force projection are, and should be duly regarded as, complementary rather than competitive in the service of joint-force commanders, since land-based bombers and fighters and carrier-based fighters are not duplicative and redundant but rather offer overlapping and mutually reinforcing as well as unique capabilities for conducting joint strike warfare (see Figure 10.1)[2].

For example, Air Force long-range bombers can penetrate deeper beyond littoral reaches than can carrier-based strike fighters supported solely by organic tanking. They also can launch directly from their home bases in the United States if no carrier strike group is positioned within immediate reach of a designated target area. Unlike bombers, however, carrier air power can provide a sustained presence as long as may be required over a target area once it is in place and provided with the requisite nonorganic tanker support.[3] The greatest liability of aircraft carriers for immediate crisis response is that they may not be close enough on short notice to where they are needed the most. In sharp contrast, the greatest advantage of long-range bombers is that they can be over a target complex anywhere in the world within 20 hours of

---

[2] This figure is a development of a most instructive graphic that originally appeared in David A. Perin, Angelyn Jewell, Barry F. McCoy, and Stephen C. Munchak, *Comparing Land-Based and Sea-Based Aircraft: Circumstances Make a Difference*, Alexandria, Va.: Center for Naval Analyses, May 1995.

[3] This point actually requires a qualification at the margins. The extent of uninterrupted airborne "presence" that a carrier can provide over a target area will be a function of many things, including the number of required daily sorties and the distance between the carrier operating area and the target area. In a revealing exercise conducted in 1997, USS *Nimitz* and her embarked air wing generated 975 simulated fixed-wing day and night combat sorties over a course of four days. A subsequent study by the Center for Naval Analyses concluded that the surge could have continued for another 12–24 more hours but not much more beyond that because of the depletion of available jet fuel and munitions, probable aircraft breakdowns, the need to conduct scheduled ship and aircraft maintenance, and eventual fatigue among aircrews and other personnel, all of which would have disrupted and, in some cases, necessitated a halt to flight operations. With two or more carriers on station, however, a continuous air presence can be maintained for an extended length of time, as was demonstrated both in Operation Enduring Freedom and in the three-week major-combat phase of Operation Iraqi Freedom. For further details on the surge exercise noted above, see Angelyn Jewell and Maureen Wigge, "Surge 97: Demonstrating the Carrier's Firepower Potential," *Proceedings*, September 1998, pp. 79–81.

Figure 10.1
Attributes of Different Forms of Air Power

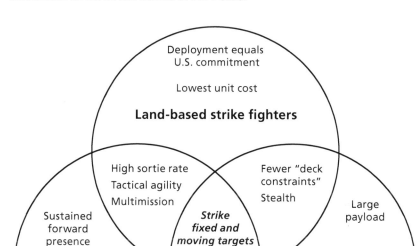

takeoff. The downside for bombers, however, is that they cannot loiter for long and regenerate striking power once their munitions have been expended, whereas carriers—especially with more than one on station—can offer persistence around the clock once they are in place.

Therein lies the synergy offered by Air Force bombers and land-based fighters and Navy carrier air wings when employed in an integrated fashion, as was amply demonstrated over Afghanistan and Iraq during the first two American wars following the terrorist attacks of September 11, 2001. As one commentator noted in this regard long before those two wars bore compelling witness to his observation, "bombers are quick to respond over vast distances to deliver very large bomb loads to an increasing variety of targets, but they are not as responsive to quick-turnaround requirements. Carrier air provides a visible presence and does not need anyone's permission to 'be there,' but has limited assets and potentially long deployment times. Theater-based attack air has the potential to provide quick turnaround in high numbers and can deploy relatively quickly but is dependent on a dwindling number of forward bases. In short, each element has strengths and weaknesses. To shortchange any one area is to hamstring the nation's ability to protect its global interests."[4]

One area in particular in which land-based and sea-based air power has a symbiotic relationship that warrants further nurturing has to do with nonorganic in-flight refueling. As was shown during Operations Enduring Freedom and Iraqi Freedom, the participating Navy carrier air wings plainly needed the support of long-range Air Force and allied tankers to generate mission-effective sorties on a sustained basis. Yet the tankers also needed the protective screening against potential enemy threats that was offered by Navy fighters in a situation in which land-based fighters were unavailable in sufficient numbers because of the lack of adequate regional basing. For his part, especially in the case of Operation Enduring Freedom over remote Afghanistan, the air component commander needed *both* force elements in order for the nation's air weapon to offer its greatest contribution to joint warfare, a fact that bore out the observation of one Air Force advocate almost a decade before that "there is a place on the team for *all* the nation's land, sea, air, and space forces," with the only real question being one of appropriate mix and affordability.[5]

---

[4] Lieutenant Colonel Gene Myers, USAF (Ret.), "Bomber Debates," *Proceedings*, August 1996, p. 35. For more on these points, see David A. Perin, *Some Observations on the Sortie Rates of Land-Based and Sea-Based Tactical Aircraft*, Alexandria, Va.: Center for Naval Analyses, March 1995; Adam B. Siegel, *Basing and Other Constraints on Land-Based Aviation Contributions to U.S. Contingency Operations*, Alexandria, Va.: Center for Naval Analyses, March 1995; and Perin and others, *Comparing Land-Based and Sea-Based Aircraft*.

[5] Colonel Brian E. Wages, USAF (Ret.), "Circle the Carriers: Why Does 'Virtual Presence' Scare the Navy?" *Armed Forces Journal International*, July 1995, p. 31, emphasis added. On the above point about force mix and affordability, it bears stressing here that the relative weight of investment that should be apportioned to each of these three force categories is a separate question altogether, and that there is no simple a priori "one-size-fits-all" solution

As attested by the Enduring Freedom experience in particular, the Navy now has every incentive to make the most of the deep-attack synergy that comes from mixing modern carrier-based strike fighters with Air Force tankers—the latter of which can *always* be counted on to be there if the prior planning is done correctly. Admiral Jay Johnson, the CNO from 1996 to 2000, essentially acknowledged this more than four years before the real-world test of Enduring Freedom when he observed that the distance that naval strike aviation reaches inland will depend on, among other things, "the contributions of joint and coalition forces." The former CNO further noted that mission needs may dictate relying on those contributions when the challenge for naval aviation is to exercise its "considerable reach and operate far inland."[6] Accordingly, whenever carrier-based aircraft are part of a joint-force employment plan within an ATO context, it should only be expected that Air Force and possibly also allied long-range tankers will be available to support Navy ATO-assigned missions.

True enough, for a number of logistical reasons having to do with forward basing limitations and resultant tanker beddown shortfalls, there were some friction points at first during Iraqi Freedom with regard to the provision of land-based tanker support to carrier-based strike fighters operating out of the eastern Mediterranean. But the Enduring Freedom experience that preceded it should have proven beyond anyone's doubt that even in a remote part of the world far from accessible shore bases, land-based tanker support will be available when needed and properly arranged for. The CFMCC during the Afghan war, Vice Admiral Moore, offered resounding testimony to this fact when he subsequently remarked in an interview: "I tip my hat to the United States Air Force and our coalition tankers, as well as our Navy's organic S-3 refueling aircraft. In cooperation with our coalition forces, we put

---

to the resource prioritization issue, since joint force commanders will mix and match forces as their assessed operational needs of the moment dictate. This report is solely about joint-service force employment at the operational and tactical levels of war. It does not consider the very different and more conflicted subject of appropriate tradeoffs between Air Force and Navy force structure.

[6] Admiral Jay L. Johnson, USN, "The Navy Operational Concept: Forward . . . from the Sea," *Sea Power*, May 1997, p. 17.

together a superb operation to provide the necessary aerial refueling capability to enable sustained tactical air operations many hundreds of miles from the sea."[7]

In both wars, to sum up, each service brought a needed comparative advantage to the fight. In the case of Enduring Freedom, Air Force bombers flew only around 10 percent of the total number of combat sorties but dropped roughly 80 percent of the ordnance, including the preponderant number of satellite-aided JDAMs. For its part, although the Navy needed the support of Air Force tankers to be mission-effective, its sea-based strike fighters operating off the coast of Pakistan from the North Arabian Sea provided an essential combat capability in a part of the world where the Air Force both lacked the needed access to operate its fighters most efficiently and remained limited in the number of fighter sorties it could generate even after it finally achieved its needed access. The reason for the latter was the substantially greater distances to Afghanistan from forward land bases in the Persian Gulf that demanded fighter missions lasting as long as 15 hours, which were unsustainable by the Air Force over the long haul.

In both cases, carrier air power, long-range bombers, land-based tankers, and land-based fighters were all eventually available and ready for CFACC tasking when the time came, and all four force elements were crucial to the timely achievement of the joint-force commander's declared objectives. Rather than continuing to engage in pointless either/or arguments over carrier vs. land-based air power that miss this overarching point, Air Force and Navy proponents should instead be using their recent combat experience as a model for seeking ways, as one writer put it nearly a decade ago, to "enhance the synergy of the air power triad of long-range projection forces" consisting of bombers, land-based fighters, and sea-based fighters that, taken together, make up the nation's overall air power equation.[8] The former commander of Naval Air Force, U.S. Atlantic Fleet, Vice Admiral John Mazach, gave clear voice to this critically important point when he reflected after the Afghan air war: "Rather than pitting one variant of air power against

---

[7] Interview with Moore, "Committed to Victory," p. 19.

[8] Myers, "Bomber Debates," p. 36.

the other . . . Enduring Freedom convincingly demonstrated that such 20th-century interservice rivalries have no place in the 21st-century U.S. warfighting establishment. The operation was remarkable for its degree of seamless interoperability between the U.S. Air Force and the Navy–Marine Corps team's sea-based aviation. . . . In short, aircraft carriers and [land-based] bombers should not be viewed as competitors for resources, but as partners able to leverage unique synergies on the modern battlefield."[9]

---

[9] Vice Admiral John Mazach, USN (Ret.), "The 21st-Century Triad: Unconventional Thinking About the New Realities of Conventional Warfare," *Sea Power*, March 2002, p. 53.

CHAPTER ELEVEN

Future Challenges and Opportunities

Air Force and Navy integration in strike warfare has shown remarkable progress in the more than three decades since the end of American combat operations in Vietnam, when such integration could be fairly said to have been almost nonexistent. By the frank admission of key participants in both services, that process of integration still has a way to go before it can be rightly described as having fully matured.[1] Nevertheless, there can be no doubt that the strike-warfare arena is now by far the most developed area of air-naval integration in the nation's joint-operations repertoire. As the most recent former commander of NSAWC, Rear Admiral Matthew Moffit, put this important point, both services are now "well into the journey" toward complete inte-

---

[1] For example, some naval aviators have recently reported that Air Force combat operations officers in the CAOC at Al Udeid Air Base in Qatar continue to be insufficiently mindful of the capability that the F/A-18E/F Super Hornet offers with its advanced-technology forward-looking infrared (ATFLIR) targeting pod and have tended to reject that aircraft repeatedly for certain targeting assignments because of a false belief that it is a "legacy" weapon system. More senior naval aviation leaders with current firsthand familiarity with CAOC operations challenge the validity of that complaint and attribute the perception behind it instead to possible other causes, such as a tendency on the part of Marine Corps ground FACs in Iraq to turn away Air Force and Navy strike fighters when Marine aircraft are available in the overhead CAS stack simply because of their greater trust in the latter. The only point of noting this example is that it attests to some continuing, if increasingly small and localized, friction points in the Air Force–Navy relationship with respect to strike warfare that beg for continuing leadership attention in both services. (Comments during a flag panel and subsequent conversations at the 2006 annual symposium of the Tailhook Association, Reno, Nevada, September 9–10, 2006.)

gration in joint strike operations.[2] Indeed, one can safely say that that journey has now progressed to a point where it can be showcased as an object lesson in the sorts of closer integration that can be successfully pursued by the Air Force and Navy in other mission areas where the air and maritime operating mediums intersect, as well as by the Air Force and Army, for that matter, when it comes to joint air-land operations.[3]

This healthy trend was not, at least at the outset, pursued entirely willfully and voluntarily by the two services. Rather, to a considerable degree, it was the cumulative outgrowth over time of a situation-driven imperative that emanated from real-world needs for the most effective employment of both services' strike assets. All the same, it most fundamentally reflected an abiding recognition and acceptance by senior leaders and line warfighters in both services that such convergence simply made good sense in the operational interests of each.

As for still-unresolved issue areas between the two services where further work can be done in the interest of closer Air Force–Navy integration, senior leaders in each service have often cited continued communications shortcomings as one important problem area in need of further attention. Within that arena, bandwidth limitations remain, by all accounts, a major constraint on the implementation of many good-in-principle ideas in the realm of C4/ISR integration that could bring the services more closely together as a joint warfighting team. One step toward a possible resolution, in the view of both Air Force and naval warfighters, would be a dynamic bandwidth management system that automatically prioritizes incoming messages.

Another persistent sore spot between the Air Force and Navy, at least from the Navy's perspective, has to do with a rapidly looming problem in the electronic attack mission area. When the Air Force decided to retire its 24 aging EF-111 Raven electronic jammer aircraft not long after Operation Desert Storm, primarily because of exces-

---

[2] Conversation with Rear Admiral Matthew Moffit, USN, Director, Fleet Readiness Division, OPNAV N43, Washington, D.C., August 1, 2006.

[3] For further discussion on this latter point, see Bruce R. Pirnie, Alan J. Vick, Adam Grissom, Karl P. Mueller, and David T. Orletsky, *Beyond Close Air Support: Forging a New Air-Ground Partnership*, Santa Monica, Calif.: RAND Corporation, MG-301-AF, 2005.

sive upkeep costs, the Navy and Marine Corps picked up the tactical electronic attack mission with their now greatly overworked EA-6B Prowlers, with the result that those aircraft became, to all intents and purposes, high-demand/low-density national assets. That arrangement has, by and large, worked satisfactorily until now, but the EA-6Bs are rapidly running out of service life, the first replacement EA-18G Growlers will not enter fleet service until 2009 at the earliest, and the interservice memorandum of agreement that made the Navy the lead service in the provision of standoff jamming after Desert Storm expires in 2011. Accordingly, senior naval aviation leaders insist that the Air Force will soon have to decide, conjointly with the Navy, what it intends to do by way of proceeding with timely gap-filler measures.[4]

For a time, the Air Force pressed to modify a number of its B-52 bombers to become long-range standoff jammers (SOJs), in response to a Department of Defense initiative aimed at adapting the B-52 for the electronic attack mission so that the Navy might begin phasing out its EA-6Bs and replacing them with the EA-18G. However, the service subsequently canceled its initial B-52 SOJ effort because the latter's cost rose from an initial estimate of $1 billion to nearly $7 billion as a result of requirements creep. As Air Force chief of staff General Moseley reflected shortly thereafter, "I think you could say that there were lots of people that took opportunities to have it be something other than the initial requirement."[5] On this issue, the Navy maintains that electronic attack is an essential key to the aircraft survivability puzzle, whereas the Air Force, in the Navy's view, has seemed willing up to now to rely more heavily on platform low observability (or "stealth"), as attested by the service's heavy investment in stealth as compared to electronic attack. As the current commander of NSAWC recently observed with respect to this issue, "both services have to figure this out. There is so much good that can be achieved by cooperating, both

---

[4] Conversation with Vice Admiral James Zortman, USN, Commander, Naval Air Forces, at the 2006 annual meeting of the Tailhook Association, Reno, Nevada, September 8, 2006, and email communication to the author from Rear Admiral Mark Emerson, USN, Commander, Naval Strike and Air Warfare Center, NAS Fallon, Nevada, January 28, 2007.

[5] Jefferson Morris, "Shifting Requirements Scuttled B-52 Stand-Off Jammer: Moseley," *Aerospace Daily and Defense Report*, April 5, 2006.

in hardware investments and TTPs, but the future bodes for a less than optimal solution (with each service appearing to go its own way), and the clock is running out for the Air Force."[6] As for the good-news part of this story, the Air Force has recently turned anew to the SOJ replacement idea, looking possibly to partner with the Navy on the EA-18G or to mount a jammer on the F-15E or a UAV—and being more mindful than ever of the need for a disciplined approach that freezes design configurations early. The Air Force also in late 2006 reopened its effort to transform some B-52s into standoff jammers through a new project redlined this time at $2 billion to protect it from the requirements growth that drove up costs and forced the cancellation of the initial B-52 SOJ program.[7]

Still other possible new or expanded joint Air Force and Navy cooperative ventures worth exploring in the realm of strike-warfare training could include, but are by no means limited to

- the development and implementation, perhaps with the close involvement of U.S. Joint Forces Command, of an improved experimentation regime aimed at extracting the greatest possible leverage from the nation's joint strike-warfare assets
- more realistic and recurrent joint peacetime training exercises between the two services as instruments for spotlighting persistent friction points, to include greater Air Force involvement in Navy carrier air wing predeployment workups at NAS Fallon and more Navy participation in Air Force Red Flag and other large-force training evolutions
- improvements in joint distributed mission simulation, which will entail major buy-in costs but which also may offer a substantial long-term payoff for both the Air Force and the Navy as fuel and associated training costs continue to soar

---

[6] Email communication to the author from Rear Admiral Mark Emerson, USN, Commander, Naval Strike and Air Warfare Center, NAS Fallon, Nevada, January 28, 2007.

[7] David A. Fulghum, "Slight of Band: Electronic Warfare Plans for the Air Force May Have a Disappearing Capability," *Aviation Week and Space Technology*, October 16, 2006, p. 38.

- a more holistic look by both services at the joint use of training ranges, especially in the southwestern portion of the United States where most strike-warfare training by the Air Force, Navy, and Marine Corps is conducted, to include further progress toward more closely linking the instrumented Nellis and Fallon range complexes
- better and more comprehensive use in a joint training setting of realistic adversary threats, not only in aerial but also in space and cyber warfare, to include possibly pooling service funds, insofar as such pooling might offer payoffs to the Air Force and Navy in equal measure
- extending integrated Air Force and Navy strike-warfare training to the surface and subsurface Navy through such measures as the inclusion in joint exercises of TLAM operations and surface-to-air fires from ships
- bringing integrated Army Patriot SAM theater air defense operations (already at Red Flag and soon to come to Fallon) into realistic Air Force and Navy joint strike-warfare training[8]
- developing and fielding compatible cross-service software and aircraft pods for fuller employment of each service's air combat maneuvering instrumentation ranges and capabilities in joint peacetime training
- making better use of networking possibilities by enlisting the real-time participation of air operations centers worldwide that

---

[8] The practical learning value to be gained from this particular proposed initiative bears further comment. In marked contrast to the increasingly seamless integration of Air Force and Navy aerial strike-warfare operations, the three incidents of fratricide involving Army Patriot SAMs and allied fixed-wing aircraft during the major combat phase of Operation Iraqi Freedom in 2003 stand as striking illustrations of some serious interoperability problems that continue to afflict the Army's relationship with its sister services at the operational and tactical levels of war. In the space of less than a week, Army Patriots in Kuwait shot down two coalition aircraft, killing three aircrew members. In addition, an Air Force F-16CJ pilot mistook a recently moved Patriot for an Iraqi SA-2 SAM and destroyed its radar with an AGM-88 high-speed antiradiation missile (HARM). For the most authoritative U.S. government–sponsored overview assessment of these incidents currently available in the public domain, see *Report of the Defense Science Board Task Force on Patriot System Performance: Report Summary*, Washington, D.C.: Office of the Assistant Secretary of Defense for Acquisition, Technology, and Logistics, January 2005.

are not focused on actual ongoing combat operations, such as those at Nellis, Langley, and Hickam AFBs, to generate realistic ATOs for large-force strike training exercises conducted at Nellis, Fallon, and elsewhere

- developing and fielding the needed wherewithal for routinely conducting geographically separated training mission debriefings by video teleconferencing rather than through telephone calls
- bringing more closely into this continuing integration process key allies such as the United Kingdom and Australia, whose current and prospective strike capabilities, other hardware assets, and known operating prowess will often offer significant value-added to U.S. strike assets in joint and combined warfare.

Many such worthy initiatives are already being cooperatively pursued, or at least carefully considered, by the Air Force Warfare Center at Nellis AFB and the Naval Strike and Air Warfare Center at NAS Fallon, with the primary limiting factor being insufficient funds to support them. As for related areas of possibly closer Air Force and Navy integration that pertain more to investments in equipment and hardware capability, the two services could usefully consider

- continued pursuit of ways to bring the Air Force's C2 Constellation and the Navy's FORCENet systems into closer horizontal integration
- greater cross-service attention to exploiting the power and promise of new electronic warfare means in a joint setting
- further cooperation and coordination among the Air Force, Navy, and Marine Corps in seeking the greatest combat leverage and operating efficiencies at the least cost from the impending introduction of the high-commonality F-35 multirole combat aircraft that will enter the inventories of all three services in the coming decade
- further joint coordination at the highest levels in determining and setting agreed-on integration priorities, since some initiatives will clearly be more important and time-urgent than others.

Finally, in the studies and analysis arena, one potentially high-payoff initiative that would cost essentially nothing beyond a determined Air Force and Navy effort to devote the right talent to it would be a careful review of any and all archived aircrew mission reports and other operational records associated with such past training exercises as Kansas Global Lancer, Roving Sands, Rugged Nautilus, Northern Edge, and Valiant Shield, as well as any and all accessible documentation pertaining to such actual contingency operations as Northern and Southern Watch and such full-fledged air-warfare experiences as Operations Deliberate Force, Allied Force, Enduring Freedom, and Iraqi Freedom, in search of any identifiable seams or friction points in integrated Air Force and Navy strike operations that were encountered along the way that may still be in need of cooperative attention and correction by both services.

In this last respect, to cite one case in point that might offer an instructive example of the kinds of challenges that can suddenly pop up at the tactical level to introduce unexpected friction in joint operations, a technical problem that arose more than once during Operation Deliberate Force in 1995 stemmed from the joint use of the Air Force's GBU-15 powered electro-optically guided 2,000-lb bomb and the Navy's SLAM in close proximity—an occurrence that produced electronic interference problems that, in the words of one Air Force assessor, "surprised everyone." This interference led to the electro-optical presentation of the GBU-15's target-area picture, normally presented on the cockpit display of the Air Force F-15E that launched the weapon, intruding on the Navy F/A-18's cockpit video display while the Hornet pilot was seeking to guide his SLAM. As a subsequent assessment noted: "This problem trashed seven SLAMs because of command-guidance failures. The costly joint-employment lesson learned here is that in future conflicts, one should write SPINs to coordinate and deconflict platforms and standoff weapons in the area, along with their respective electro-optical frequency spectrums."[9] A searching joint inventory and assessment of other such challenges, both real and potential, by experts at the Air Force's and Navy's air warfare centers

---

[9] Sargent, "Deliberate Force Tactics," in Owen, *Deliberate Force*, p. 324.

at Nellis and Fallon or elsewhere might prove to be both a worthwhile and low-cost investment of time in the interest of moving both services one more step toward achieving a fully mature integrated strike-warfare capability.

Even with this much room remaining for further progress by the two services, however, the overall record of Air Force and Navy accomplishment in integrated air warfare planning and operations throughout the more than three decades since Vietnam stands as a resounding good-news story that is a credit to each service both separately and together. A central theme woven throughout the preceding pages has been that Air Force and Navy strike-warfare capabilities and repertoires have, since Vietnam, become almost seamlessly integrated in a way, and to an extent, that cannot yet be said of any other two U.S. force elements. As such, they represent a role model for what can be done along similar lines elsewhere, not just in the interface between air and maritime operations, but even more so in the still-troubled relationship between the Air Force and Army when it comes to the most efficient conduct of joint air-land warfare.[10]

In hindsight, those naval aviation leaders who were "present at the creation" of this remarkable process would probably agree that at the outset, at least, it was the Navy in the early aftermath of Operation Desert Storm that was obliged to do most of the accommodating, since it was, by the frank admission of both its leaders and its working-level operators at the time, the service that faced the greater integration needs because of its unique configuration and posturing for a challenge that, almost overnight, had ceased to exist with the disappearance of the Soviet threat. Accordingly, it almost naturally had greater initial incentives to change in order to remain relevant in the rapidly emerging post–Cold War world. As the current commander of NSAWC recently observed on this point, "the Navy was disadvantaged, to be sure—it had to make more significant changes and adaptations, and Desert

---

[10] For a thoughtful amplification in depth on this broad and still-unresolved issue area, see David E. Johnson, *Learning Large Lessons: The Evolving Roles of Ground Power and Air Power in the Post–Cold War Era*, Santa Monica, Calif.: RAND Corporation, MG-405-AF, 2006.

Storm provided the event that forced us to recognize it clearly."[11] Yet at the same time, the Air Force also found itself compelled to abandon many of its parochial fixations and habits that had accumulated over decades of confronting a very different set of operating challenges, perhaps most notably its deep-seated proclivity, as perceived by the naval air warfare community at the time, of insisting that "you'll do it the USAF way, period," which did not exactly engender a positive working relationship at the outset.[12] That said, once the initial hurdles were crossed and the two services found themselves increasingly working from a common playbook, the bulk of evidence presented in the foregoing discussion suggests that each brought a similarly solutions-oriented perspective to the table whenever significant cross-service differences in TTPs arose that needed to be ironed out and adjudicated.

Indeed, the closeness of joint-service integration in the air warfare arena that right-thinking leaders in both services have progressively nurtured in more recent years should be both applauded as a collective Air Force and Navy accomplishment and strongly encouraged on the part of other warfighting communities. Furthermore, although the foregoing discussion may have conveyed an impression that the process was both easy and natural, in fact the operational integration described in this study had to overcome multiple barriers and the most deeply ingrained resistance to change in both services. The fact that organizations, especially military organizations, tend to resist rather than embrace change makes the history and experience described above all the more remarkable.[13] That history and experience represent

---

[11] Email communication to the author from Rear Admiral Mark Emerson, USN, Commander, Naval Strike and Air Warfare Center, NAS Fallon, Nevada, January 28, 2007.

[12] Email communication to the author from Rear Admiral Mark Emerson. To cite just one example provided in the recollections of Admiral Emerson, who served during Desert Storm as an F/A-18 pilot and as the operations officer of Carrier Air Wing 5 deployed in USS *Midway*: "Many naval/littoral targets were not being given adequate priority" by the Air Force–dominated CAOC, such as "Iraqi naval mines—we could have reduced their effectiveness had we had the opportunity to destroy them in warehouses rather than spending huge amounts of time clearing them later."

[13] I am indebted to my RAND colleague John Stillion for bringing this important point to my attention.

a true object lesson in how joint-force integration should take place as a matter of practice.

Even more encouraging, thanks to the guiding role played by strong-willed and solutions-oriented individuals in both services with the right focus and a determination to act on it, there is now a well-ensconced successor generation in place in both who grew up as line aircrew members during the formative years of this integration process. Those individuals have since migrated through such mid-level positions as CAOC night coordinators, combat plans and operations staffers, and strategy division principals to the more senior flag ranks and leadership positions that will help them ensure that the strike-warfare communities in both services continue to pursue an increasingly common operational culture, as well as remain duly joint in their outlook when it comes to more effective cross-service warfighting. Today, such commonality of purpose at the operational and tactical levels has become more important than ever as the nation finds itself increasingly reliant on the combined-arms potential that is now available in principle to all services for continuing to prosecute the global war against terror, while hedging also against future peer or near-peer competitors at a time of almost unprecedented lows in annual spending for force modernization.

Bibliography

After-Action Report, "28 WG/CAG-1 Joint Fallon Strike," December 31, 1991, cited in Captain James W. Fryer, USAF, "Flying with the Bone," *Proceedings*, February 1995.

After-Action Report on Kansas Global Lancer, "Joint USN-USAF B-1B Strike Postex," Command Sixth Fleet, April 10, 1993, cited in Captain James W. Fryer, USAF, "Flying with the Bone," *Proceedings*, February 1995.

"Air Force Chief Says Munition Stockpiles Are Sound," *Inside the Air Force*, April 11, 2003.

Arkin, William M., "Old-Timers Prove Invaluable in Afghanistan Air Campaign," *Los Angeles Times*, February 10, 2002.

Baker, Peter, and Rajiv Chandrasekaran, "Republican Guard Units Move South from Baghdad, Hit by U.S. Forces," *Washington Post*, March 27, 2003.

Barry, Major General John L., USAF, and James Blaker, "After the Storm: The Growing Convergence of the Air Force and Navy," *Naval War College Review*, Autumn 2001.

Bennett, John T., "USAF Constellation, Navy FORCENet Step Up Interoperability Efforts," *Inside the Pentagon*, July 31, 2003.

Bien, Captain Lyle G., USN, "From the Strike Cell," *Proceedings*, June 1991.

Burgess, Richard R., "Air Strikes Hit Afghan Front Lines," *Sea Power*, December 2001.

Burns, Robert, "U.S. Gulf Force Nears 300,000 as Commander, Bush Consult," *Philadelphia Inquirer*, March 5, 2003.

Campbell, Colonel Christopher M., USAF, "The Deliberate Force Air Campaign Plan," cited in Colonel Robert C. Owen, USAF, *Deliberate Force: A Case Study in Effective Air Campaigning*, Maxwell AFB, Ala.: Air University Press, 1999.

Cebrowski, Vice Admiral Arthur K., USN, and John J. Garstka, "Network-Centric Warfare: Its Origin and Future," *Proceedings,* January 1998.

Conversino, Mark, "Executing Deliberate Force: 30 August–14 September 1995," cited in Colonel Robert C. Owen, USAF, *Deliberate Force: A Case Study in Effective Air Campaigning*, Maxwell AFB, Ala.: Air University Press, 1999.

Crawley, Vince, "Less Is More," *Army Times*, April 21, 2003.

Dalton, John, Admiral Jeremy Boorda, and General Carl Mundy, Jr., "Forward . . . from the Sea," *Proceedings*, December 1994.

Davidson, Lieutenant General Phillip B., USA (Ret.), *Vietnam at War: The History, 1946–1975*, New York: Oxford University Press, 1988.

Department of Defense Dictionary of Military and Associated Terms, Washington, D.C.: Department of Defense, Joint Publication 1-02, April 12, 2006.

Di Rita, Lieutenant Commander Larry, USN, "Exocets, Air Traffic, and the Air Tasking Order," *Proceedings*, August 1992.

Dowdy, William L., . . . *From the Sea: Preparing the Naval Service for the 21st Century*, Washington, D.C.: Department of the Navy, 1992.

———, *Testing the Aerospace Expeditionary Force Concept: An Analysis of AEF's I–IV (1995–97) and the Way Ahead*, Maxwell AFB, Ala.: College of Aerospace Doctrine, Research, and Education, Air University, Research Paper 2000–01, 2000.

Faletti, Lieutenant Commander Matthew J., USN, "Close Air Support Must Be Joint," *Proceedings*, September 1994.

"Fitzgerald: Recapitalization Poses Challenge for Naval Air," *Sea Power*, March 2004.

Fryer, Captain James W., USAF, "Flying with the Bone," *Proceedings*, February 1995.

Fulghum, David A., "Slight of Band: Electronic Warfare Plans for the Air Force May Have a Disappearing Capability," *Aviation Week and Space Technology*, October 16, 2006.

Gandt, Robert, *Bogeys and Bandits: The Making of a Fighter Pilot*, New York: Penguin Books, 1997.

Garcia-Barham, Lieutenant Denise, USN, "U.S. Warships Complete Exercise in Alaska," *Navy NewsStand*, August 26, 2005.

Global Reach, Global Power, White Paper, Washington, D.C.: Department of the Air Force, December 1992.

Grant, Rebecca, "The War Nobody Expected," *Air Force Magazine*, April 2002.

Hardesty, Captain David C., USN, "Fix Net Centric for the Operators," *Proceedings*, September 2003.

Hehs, Eric, "NSAWC," *Code One*, Fort Worth, Tex.: Lockheed Martin Aeronautics Corporation, October 1998.

Hewish, Mark, "Out of CAOCs Comes Order," *Jane's International Defense Review*, May 2003.

Hunt, Peter, *Angles of Attack: An A-6 Intruder Pilot's War*, New York: Ballantine Books, 2002.

Jewell, Angelyn, and Maureen Wigge, "Surge 97: Demonstrating the Carrier's Firepower Potential," *Proceedings*, September 1998.

Johnson, David E., *Learning Large Lessons: The Evolving Roles of Ground Power and Air Power in the Post–Cold War Era*, Santa Monica, Calif.: RAND Corporation, MG-405-1-AF, 2006. As of August 1, 2007:
http://www.rand.org/pubs/monographs/MG405-1/

Johnson, Admiral Jay L., USN, "The Navy Operational Concept: Forward . . . from the Sea," *Sea Power*, May 1997.

Kauchak, Marty, "Navigating Changing Seas: Navy Chief Harbors No Illusions About the Challenges That Lie Ahead," *Armed Forces Journal International,* August 2002.

Keating, Vice Admiral Timothy J., USN, interview, "This Was a Different War," *Proceedings*, June 2003.

Keeter, Hunter, "Navy's Lessons from Afghanistan, Iraq Include Networked Tactical Aircraft," *Defense Daily*, May 6, 2003.

Knights, Michael, *Cradle of Conflict: Iraq and the Birth of the Modern U.S. Military*, Annapolis, Md.: Naval Institute Press, 2005.

Lambeth, Benjamin S., *The Transformation of American Air Power*, Ithaca, N.Y.: Cornell University Press, 2000.

———, *NATO's Air War for Kosovo: A Strategic and Operational Assessment*, Santa Monica, Calif.: RAND Corporation, MR-1365-AF, 2001. As of August 1, 2007: http://www.rand.org/pubs/monograph_reports/MR1365/

———, *Air Power Against Terror: America's Conduct of Operation Enduring Freedom*, Santa Monica, Calif.: RAND Corporation, MG-166-1-CENTAF, 2005. As of August 1, 2007:
http://www.rand.org/pubs/monographs/MG166-1/

———, *American Carrier Air Power at the Dawn of a New Century*, Santa Monica, Calif.: RAND Corporation, MG-404-NAVY, 2005. As of August 1, 2007: http://www.rand.org/pubs/monographs/MG404/

Leopold, Reuven, *Sea-Based Aviation and the Next U.S. Aircraft Carrier Design: The CVX*, MIT Security Studies Program Occasional Paper, Cambridge, Mass.: Center for International Studies, Massachusetts Institute of Technology, January 1998.

Levardi, Captain Yvonne, USAF, "Air Ops Center Wraps Up Valiant Shield," news release, Hickam AFB, Hawaii: Office of Public Affairs, Kenney Warfighting Headquarters, June 26, 2006.

Lockard, Vice Admiral John A., USN, interview, "Affordable Readiness for the Operator: 21st-Century Aviation Solutions Enable Dominance from the Sea," *Sea Power*, June 1999.

Malone, Vice Admiral Michael, USN, "They Made a Difference," *The Hook*, Summer 2003.

Marolda, Edward J., and Robert J. Schneider, Jr., *Sword and Shield: The United States Navy and the Persian Gulf War*, Annapolis, Md.: Naval Institute Press, 1998.

Mazach, Vice Admiral John, USN (Ret.), "The 21st-Century Triad: Unconventional Thinking About the New Realities of Conventional Warfare," *Sea Power*, March 2002.

McNamara, Lieutenant Colonel Stephen J., USAF, *Air Power's Gordian Knot: Centralized versus Organic Control*, Maxwell AFB, Ala.: Air University Press, 1994.

Mixon, Rear Admiral Riley D., USN, "Where We Must Do Better," *Proceedings*, August 1991.

Moore, Vice Admiral Charles W., Jr., USN, former Commander, U.S. Naval Forces Central Command and Commander, Fifth Fleet, interview, "Committed to Victory," *Sea Power*, March 2002

Moore, Commander Daniel E., USN, "Bosnia, Tanks, and 'From the Sea,'" *Proceedings*, December 1994.

Moore, Lieutenant General Royal N., Jr., USMC, "Marine Air: There When Needed," *Proceedings*, November 1991.

Morris, Jefferson, "Shifting Requirements Scuttled B-52 Stand-Off Jammer: Moseley," *Aerospace Daily and Defense Report*, April 5, 2006.

Morton, John F., "The U.S. Navy in 1991," *Proceedings*, Naval Review 1992.

Moseley, Lieutenant General T. Michael, USAF, *Operation Iraqi Freedom—By the Numbers*, Shaw AFB, S.C.: Assessment and Analysis Division, Headquarters U.S. Central Command Air Forces, April 30, 2003.

Muir, Commander Daniel J., USN, "A View from the Black Hole," *Proceedings*, October 1991.

Murphy, Kim, and Alan C. Miller, "The Team That Picks the Targets," *Los Angeles Times*, March 25, 2003.

Murray, Williamson, and Major General Robert H. Scales, Jr., USA (Ret.), *The Iraq War: A Military History*, Cambridge, Mass.: The Belknap Press of Harvard University Press, 2003.

Myers, Lieutenant Colonel Gene, USAF (Ret.), "Bomber Debates," *Proceedings*, August 1996.

Natter, Admiral Robert J., "New Command Unifies the Fleet," *Proceedings,* January 2002.

"Navy to Boost Image, Relations with Press," *Aviation Week and Space Technology*, September 23, 1991.

Oliver, Commander J. D., III, USN, "To Train to Fight," *Proceedings*, September 1995.

Owen, Colonel Robert C., USAF, *Deliberate Force: A Case Study in Effective Air Campaigning*, Maxwell AFB, Ala.: Air University Press, 1999.

Owens, Vice Admiral William A., USN, "The Quest for Consensus," *Proceedings*, May 1994.

Owens, Admiral William A., USN (Ret.), *High Seas: The Naval Passage to an Uncharted World*, Annapolis, Md.: Naval Institute Press, 1995.

Palzkill, Lieutenant Dennis, USN, "Making Interoperability Work," *Proceedings*, September 1991.

Pan, Philip P., "Turkish Leader Makes Request on Airspace," *Washington Post*, March 20, 2003.

Perin, David A., *Some Observations on the Sortie Rates of Land-Based and Sea-Based Tactical Aircraft*, Alexandria, Va.: Center for Naval Analyses, March 1995.

Perin, David A., Angelyn Jewell, Barry F. McCoy, and Stephen C. Munchak, *Comparing Land-Based and Sea-Based Aircraft: Circumstances Make a Difference*, Alexandria, Va.: Center for Naval Analyses, May 1995.

Peterson, Gordon I., "Naval Aviation Spearheads Operation Noble Anvil," *Sea Power*, June 1999.

Pirnie, Bruce R., Alan J. Vick, Adam Grissom, Karl P. Mueller, and David T. Orletsky, *Beyond Close Air Support: Forging a New Air-Ground Partnership*, Santa Monica, Calif.: RAND Corporation, MG-301-AF, 2005. As of August 1, 2007: http://www.rand.org/pubs/monographs/MG301/

Pollack, Kevin E., "Desert Storm Taught Us Something," *Proceedings*, January 1995.

Redd, Vice Admiral Scott, USN (Ret.), "Sustained Assured Access: The Navy's Role in Joint Warfare," *Armed Forces Journal International*, September 2001.

Report of the Defense Science Board Task Force on Patriot System Performance: Report Summary, Washington, D.C.: Office of the Assistant Secretary of Defense for Acquisition, Technology, and Logistics, January 2005.

Ricks, Thomas E., "What Counted: People, Plan, Inept Enemy," *Washington Post*, April 10, 2003.

Ricks, Thomas E., and Vernon Loeb, "Initial Aim Is Hitting Taliban Defenses," *Washington Post*, October 8, 2001.

Roos, John G., "Turning Up the Heat: Taliban Became Firm Believers in Effects-Based Operations," *Armed Forces Journal International,* February 2002.

Rowe, Lieutenant Commander Steve, USNR, "Saving Naval Aviation," *Proceedings*, September 2000.

Sanger, David E., with Warren Hoge, "U.S. May Abandon UN Vote on Iraq, Powell Testifies," *New York Times*, March 14, 2003.

Sargent, Lieutenant Colonel Richard L., USAF, "Deliberate Force Combat Air Assessments," cited in Colonel Robert C. Owen, USAF, *Deliberate Force: A Case Study in Effective Air Campaigning*, Maxwell AFB, Ala.: Air University Press, 1999.

Schank, John, Harry Thie, Clifford Graf, Joseph Beel, and Jerry Solinger, *Finding the Right Balance: Simulator and Live Training for Navy Units*, Santa Monica, Calif.: RAND Corporation, MR-1441-NAVY, 2002. As of August 1, 2007: http://www.rand.org/pubs/monograph_reports/MR1441/

Schmitt, Eric, "Pentagon Ready to Strike Iraq Within Days if Bush Gives the Word, Officials Say," *New York Times,* March 6, 2003.

Scott, William B., "Fallon Becoming Navy's Air Combat 'Grad School,'" *Aviation Week and Space Technology*, March 8, 1999.

Shanker, Thom, and Eric Schmitt, "Rumsfeld Seeks Consensus Through Jousting," *New York Times*, March 19, 2003.

Siegel, Adam B., *Basing and Other Constraints on Land-Based Aviation Contributions to U.S. Contingency Operations*, Alexandria, Va.: Center for Naval Analyses, March 1995.

Smith, Captain Edward A., Jr., USN, "What '. . . From the Sea' Didn't Say," *Naval War College Review*, Winter 1995.

Stephens, Hampton, "Air Force, Sister Services Begin Work to Link Network Warfare Visions," *Inside the Air Force*, February 27, 2004.

Strike Fighter Squadron 87, "Aircraft—Yes, Tactics—Yes, Weapons—No," *Proceedings*, September 1991.

Truver, Scott C., "The U.S. Navy in Review," *Proceedings*, May 2003.

Tuck, Shane, "Valiant Shield Provides Valuable Joint Training Among U.S. Military Forces, *Navy Newsstand*, June 20, 2006.

"USS *Theodore Roosevelt* Pounds Taliban and Terrorist Targets," *Sea Power*, December 2001.

Vogel, Steve, "Gas Stations in the Sky Extend Fighters' Reach," *Washington Post*, November 1, 2001.

Wages, Colonel Brian E., USAF (Ret.), "Circle the Carriers: Why Does 'Virtual Presence' Scare the Navy?" *Armed Forces Journal International*, July 1995.

Wall, Robert, "Navy Adapts Operations for Afghan War Hurdles," *Aviation Week and Space Technology*, November 19, 2001.

————, "Battle Management Dominates E-2C Combat Operations," *Aviation Week and Space Technology*, November 26, 2001.

Walsh, Edward J., "The Copernican Revolution: U.S. Navy Leads the C4I Way for Joint Operations," *Armed Forces Journal International*, July 1995.

Winnefeld, Commander James A., Jr., USN, "It's Time for a Revival," *Proceedings*, September 1992.

Winnefeld, James A., and Dana J. Johnson, *Joint Air Operations: Pursuit of Unity in Command and Control, 1942–1991*, Annapolis, Md.: Naval Institute Press, 1993.